STO

/12713

Analytical Chemistry of Liquid Fuel Sources

Analytical Chemistry of Liquid Fuel Sources

Tar Sands, Oil Shale, Coal, and Petroleum

**Peter C. Uden and
Sidney Siggia,** EDITORS
University of Massachusetts

Howard B. Jensen, EDITOR
Laramie Energy Research Center

Based on a symposium cosponsored

by the Divisions of Petroleum Chemistry

and Analytical Chemistry at the 173rd

Meeting of the American Chemical

Society, New Orleans, Louisiana

March 21–25, 1977.

ADVANCES IN CHEMISTRY SERIES **170**

AMERICAN CHEMICAL SOCIETY

WASHINGTON, D. C.   1978

Library of Congress CIP Data

Analytical chemistry of liquid fuel sources.
  (ACS advances in chemistry series; 170 ISSN 0065–2393)

  Papers presented at a symposium held in Mar. 1977 at New Orleans, sponsored by the Division of Petroleum Chemistry and the Division of Analytical Chemistry of the American Chemical Society.
  Includes bibliographies and index.

  1. Liquid fuels—Congresses.
  I. Uden, Peter C., 1939– . II. Jensen, Howard Barnett, 1921– . III. Siggia, Sidney. IV. American Chemical Society. Division of Petroleum Chemistry. V. American Chemical Society. Division of Analytical Chemistry. VI. Series: Advances in chemistry series; 170.

QD1.A355 no. 170   [TP343] 540'.8s [662'.6]
                                    78–10399
ISBN 0–8412–0395–4   ADCSAJ 170 1–342 1978

Advances in Chemistry Series

Robert F. Gould, *Editor*

Advisory Board

FOREWORD

ADVANCES IN CHEMISTRY SERIES was founded in 1949 by the American Chemical Society as an outlet for symposia and collections of data in special areas of topical interest that could not be accommodated in the Society's journals. It provides a medium for symposia that would otherwise be fragmented, their papers, distributed among several journals or not published at all. Papers are reviewed critically according to ACS editorial standards and receive the careful attention and processing characteristic of ACS publications. Volumes in the ADVANCES IN CHEMISTRY SERIES maintain the integrity of the symposia on which they are based; however, verbatim reproductions of previously published papers are not accepted. Papers may include reports of research as well as reviews since symposia may embrace both types of presentation.

CONTENTS

PREFACE

The role of analytical chemistry in the broadening field of liquid fuel technology increasingly exemplifies the latest advances in technique and instrumentation. Indeed, so complex is the area that the whole range of presently available methods for quantitation and characterization are needed in its study.

Over the past few years, established analytical chemical methodology for crude oil and refined petroleum derivatives has been extended to the rapidly expanding field of coal liquefaction products and has assisted in the substantive reappraisal of such potential liquid fuel sources as oil shale, tar sands, and similar bituminous deposits. While many of the analytical problems of separation, identification, and characterization are common to all of these fields, each area exhibits distinct requirements calling for specific development of appropriate methodology. Indeed, the added chemical complexity of the nonpetroleum-based liquid fuel sources presents many novel challenges to the chemical investigator.

In order to bring together present expertise in the various areas of liquid fuel analysis and characterization, the ACS Divisions of Petroleum Chemistry and Analytical Chemistry cosponsored a symposium wherein papers presented both reviewed general and specific analytical methodology of particular liquid fuel sources and also introduced applications of many currently developing techniques. Methods ranged from high-resolution gas and liquid chromatography to electron microprobe, carbon-13 NMR, EPR, and computer modeling. Efforts were made to indicate where techniques developed in one fuel liquids area could be applied usefully in others.

Many of the papers presented at that symposium have been revised and updated for the present volume and we thank all the contributors for their cooperation in this project. We are also grateful to the ACS Divisions of Petroleum Chemistry and Analytical Chemistry for their support.

HOWARD B. JENSEN
United States Department of Energy
Laramie Energy Research Center
Laramie, WY
June 30, 1978

SIDNEY SIGGIA and
PETER C. UDEN
Department of Chemistry
University of Massachusetts
Amherst, MA

Spontaneous Combustion Liability of Subbituminous Coals: Development of a Simplified Test Method for Field Lab/Mine Applications

WILLIAM A. SCHMELING, JEANNETTE KING,
and JOSEF J. SCHMIDT–COLLERUS

Denver Research Institute, University of Denver, Denver, CO 80208

Eighteen subbituminous coal samples were tested for their liability to spontaneous combustion using a combination of temperature-programmed air oxidation and gas–liquid chromatography, the latter being used in lieu of the conventional Orsat apparatus. Various column and switching arrangements as well as operational parameters were investigated with emphasis on simplicity and the ultimate objective to develop a portable instrument which can be used effectively for on-site testing or in field laboratories. Such an instrument could be utilized with fresh samples, thus preventing the usual sample degradation. Correlations of gas concentrations as a function of temperature and the S-index (which reflects the liability of the coal to spontaneous combustion) were determined by this method. The values obtained and the evaluation of the effectiveness of the tested analytical method are discussed.

The conservation of our fossil fuel resources is of great concern in these times of energy shortages. This is true for coal, especially in view of the eminent development of large-scale synfuel production from this resource. The danger of spontaneous oxidation and heating is a constant threat during mining, within large raw material storage bins for gasification or liquefaction operations, and/or in mine tailings. In order to minimize losses of coals, machinery, and even personnel, it is expedient

0-8412-0395-4/78/33-170-001$05.00/1

to detect spontaneous self-heating liabilities in coals as early as possible so that timely remedial action may be taken.

Since coals release gaseous products, it was thought that this property might be used as an indicator of a coal's susceptibility to spontaneous oxidation. Probably the most systematic research in this area was carried out by Winmill and Graham (1) in a special laboratory for the study of spontaneous coal combustion operated by the Doncaster Coal Owners Association. These investigations were published in a series of papers in the "Transactions of the Institution of Mining Engineers" between 1913 and 1935. The tangible results of this work were the development of a method using the evolution of carbon monoxide, in combination with the determination of oxygen deficiency, for the evaluation of the spontaneous heating liability of coals. This method has now been in general use for about forty years. However, more recent developments of sophisticated gas analysis equipment has all but replaced the classical Orsat method (2, 3). In particular, gas chromatography is capable of analyzing all constituents associated with coal oxidation to a high degree of accuracy. Furthermore, some GC models are portable and can be used in the field operated by personnel without extensive training.

Besides measuring gaseous effluents directly at the mine site, laboratory tests have been designed for evaluating a coal's liability to spontaneous heating. Finely ground samples are heated either adiabatically or nonadiabatically while being exposed to a flow of moist air. The effluent gas concentrations, particularly carbon monoxide, carbon dioxide, and oxygen are indicative of a coal's liability to spontaneous combustion. Using this method, coal samples may be classified according to an S-index (susceptibility) (4) which categorizes them as either very dangerous, not very dangerous, or not dangerous with respect to spontaneous heating.

This paper describes a nonadiabatic method for the evaluation of eighteen subbituminous coal samples using the S-index values. Laboratory methods for conducting the coal oxidation as well as the analysis of the effluent gases by gas chromatography are discussed. The objective of these experiments was to develop a methodology which ultimately might be amenable to the construction of a portable instrument usable in the field or in a mine lab. This could reduce sample handling and produce fairly accurate results in a considerably shorter period of time.

Experimental

Coal samples originating from a Wyoming subbituminous coal field varied in size from fines to 1 cm^3 in diameter and were received in airtight plastic containers. The samples, as received, were ground to 0.3–0.5-mm size in a glove box under a slightly pressurized nitrogen atmosphere. To accomplish this, the coal was ground initially with a mortar and pestle. Final size reduction was carried out with a Waring blender run for short periods to minimize the production of fines. Then 100-g portions were stored in plastic bags in a nitrogen atmosphere.

The oxygen absorption apparatus was of a nonadiabatic type as described by Chamberlain, et al. (5). A schematic of the apparatus is shown in Figure 1.

Provision was made to either run air or nitrogen through the system. The latter was used for purging purposes prior to an experimental run. Air (60 mL/min) from a compressed tank was filtered to remove impurities and bubbled through water to attain 90–95% relative humidity. It was heated then in an oil bath by passing through a glass coil in order to heat it to the temperature of the coal. After reacting with the coal, the air and effluent gases passed through a moisture trap before entering the gas chromatograph. Flow rates were monitored on either side of the coal reaction chamber. A programmed temperature controller (Model West), in conjunction with a 5-W immersion heater, supplied heat to the oil bath at a rate of 25°C per hr. The temperatures of the coal and oil bath were measured with chrome–alumel thermocouples connected to a potentiometer. The 100-g coal samples were heated to 200°C or higher over approximately an 8-hr period. Oil bath and coal temperatures were recorded every 30 min and an effluent gas analysis was run on a Beckman GC–4 gas-chromatograph.

In this method, gas chromatography was used for analyzing the effluent gases because it is rapid and sensitive to low concentrations of the constituent gases. Suitable adsorption column configurations and operational parameters (adsorbent chemicals, temperatures, and carrier-gas flow rate) had to be developed. Since the objective was to use a relatively simple system that might be adaptable to field methods, it meant a reasonable carrier-gas flow rate (20–60 mL/min) and relatively low temperatures of operation (isothermal if possible). Many methods for separating mixtures containing oxygen, nitrogen, methane, carbon monoxide, carbon dioxide, ethylene, and ethane have been described in the literature (6, 7, 8, 9). These involve connecting adsorption columns for light components (O_2, N_2, CH_4, CO) with partition columns for the remaining gases. The gases of concern in the coal self-oxidation experiments are O_2, CH_4, CO, and CO_2. Reported methods for separation of these gases use (with some variations) two columns in series. The first column consists of ¼-in. tubing, 4–10 ft long, and is filled with an adsorbent to remove CO_2 and separate it from the remainder of the constituents. The adsorbents generally used are silica gel, Chromosorb 102, or Poropak Q. The second column (of more or less the same size tubing) is capable of separating the permanent gases and light hydrocarbons. Molecular sieve adsorbents such as 13X or 5A are usually used for this purpose. The objective is to allow all but the CO_2 to pass into the second column, then switch to a bypass position and allow the CO_2 to elute directly to the detector while the remainder of the gases are held in the second column. After the CO_2 is eluted completely, the system is switched back to the series position and the remaining gases (O_2, CH_4, and CO) are separated and passed to the detector. Reported temperatures (isothermal or programmed) and carrier-gas flow rates vary to some extent but otherwise the methods are similar.

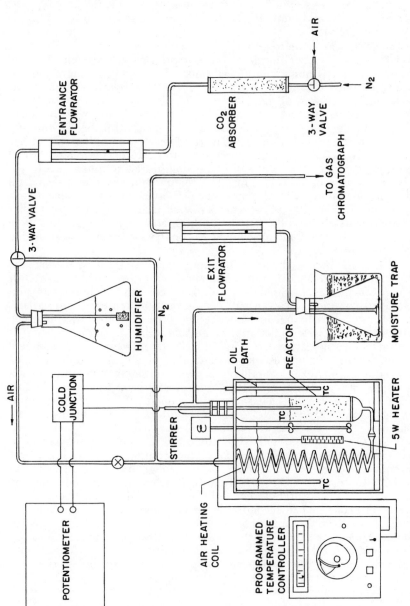

Figure 1. Nonadiabatic apparatus for determining spontaneous combustion liability of coal

Initial experiments utilized low carrier-gas flow rates and temperatures in order to ensure a sufficient retention of CO_2 in the first column (Chromosorb 102). This was accomplished to some extent; however, there was also some adsorption of CH_4, and it was difficult to switch to the bypass position without partial or total loss of either CO_2 or CH_4. On the other hand, if sufficient time was allowed for CH_4 to elute from Column 1 before switching to the bypass position, part of the CO_2 was also eluted, yielding only a partial detection as shown in Figure 2a. The

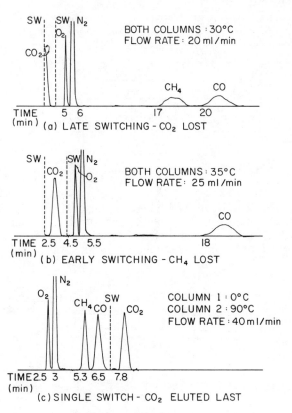

Figure 2. *Chromatograms from various column-switching configurations*

best results which could be determined under these operating conditions involved unreasonably long residence times for CH_4 and CO, and produced flat, wide peaks. If the bypass switching was carried out in time to yield a well-defined CO_2 peak, CH_4 could be missed entirely (*see* Figure 2b). Using shorter columns and higher flow rates, oxygen had begun eluting from Column 2 after switching back to series position following CO_2 elution from Column 1. Juggling of temperatures, flow rates, and column lengths and adsorption materials produced all of the peaks; however, switching had to be very precise. In view of these

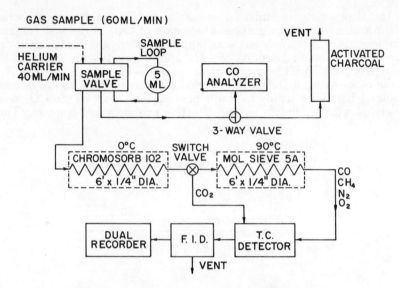

Figure 3. Gas chromatograph analysis system

problems, a scheme was devised whereby the CO_2 retention time in the first column was increased to a point that all of the other constituents eluted from the second column before CO_2 eluted from the first column. By this method only one switching operation was required, and by

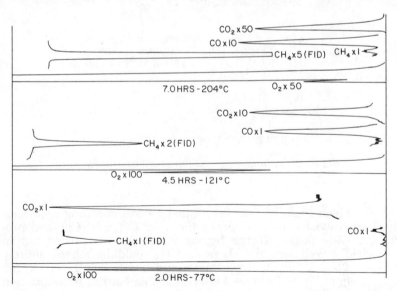

Figure 4. Chromatograms produced from the oxygen absorption of Coal No. 5

adjusting temperatures and flow rate, a reasonably short analysis time could be attained (*see* Figure 2c). Column temperatures were maintained at 0°C and 90°C, respectively, with a helium-gas flow rate of 40 mL/min.

A schematic of the gas chromatographic analysis system is shown in Figure 3. A 5-mL sampling loop was used in conjunction with the sampling valve shown. During the first stages of oxidation, carbon monoxide was monitored with an Ecolyzer CO analyzer. The residual gases were adsorbed ultimately on to an activated charcoal filter.

Oxygen, carbon monoxide, and carbon dioxide were analyzed using a thermal conductivity detector while methane was analyzed with a flame ionization detector. Chromatograms were recorded on a Varian G–2000 dual pen recorder.

Using the modified procedure described above, chromatograms were obtained every 30 min for the four constituent gases. A typical chromatogram produced for Coal No. 5 at 2 hr, 4.5 hr, and 7 hr is illustrated in Figure 4. Corresponding temperatures are indicated along with recorder attenuations used for the different gases. The increase in carbon monoxide and carbon dioxide accompanied by a decrease in oxygen with increasing temperature is readily apparent.

Calibration of the gases was carried out by means of comparison with a special Scott Calibration Mixture. Peak areas were obtained from products of the heights and widths at mid-height.

Results

The spontaneous combustion liability S-index was determined for the eighteen coal samples using the following equation:

$$S = \frac{h_1 + h_2}{2} (x_1) + \frac{h_2 + h_3}{2} (x_2) \tag{1}$$

where h_1 = oxygen deficiency at 125°C, h_2 = oxygen deficiency at 150°C, h_3 = oxygen deficiency at 175°C, x_1 = increase of CO_2 between 125° and 150°C, and x_2 = increase of CO_2 between 150° and 175°C. The oxygen deficiency was determined from the difference between the initial and final oxygen concentration. All values were reported as volume percentages.

S-Indices for various coal samples are tabulated (together with the exhaust gas analysis at 150° and 175°C) in Tables I and II. As can be observed, the S-index values range from ca. 13–47. These values represent a rather broad liability range. Values greater than 30 are considered to be in the dangerous category. The increase in carbon monoxide and carbon dioxide along with a decrease in oxygen with an increase in S-index is also obvious; however, it is much more pronounced at 175°C. Maximum gas concentrations for carbon monoxide and carbon dioxide in the effluent gases obtained at temperatures in the vicinity of 200°C were

Table I. Summary of S-Indices and Exhaust Gas Analyses at 150°C

S-Index	Coal No.	Exhaust Gas Analysis (Vol %)			
		O_2	CO_2	CO	CH_4
2.42[a]	6	12.25	0.74	.26	[b]
12.81	10	14.00	0.98	.43	.0005
13.31	7	11.50	1.61	.69	.0046
15.82	5	13.25	2.50	.57	.0032
16.75	3	12.85	1.51	.51	.0036
18.75	12	15.00	1.42	.28	[b]
20.66	2	12.90	1.95	.59	.0032
22.19	16	10.85	1.64	.94	[b]
22.85	11	11.25	1.68	.52	[b]
24.35	4	9.50	2.00	.73	.0042
25.84	14	10.75	1.91	.73	.0044
27.93	15	13.50	1.72	.56	.0025
28.34	13	11.40	1.67	.63	.0052
28.66	18	11.25	2.18	.70	.0037
33.93	17	12.25	2.50	.71	[b]
34.51	1	9.50	1.92	.67	.0002
41.00	9	10.75	1.81	.67	.0053
47.32	8	10.24	2.32	.65	.0042

[a] Low sample weight.
[b] No measurement obtained.

Table II. Summary of S-Indices and Exhaust Gas Analyses at 175°C

S-Index	Coal No.	Exhaust Gas Analysis (Vol %)			
		O_2	CO_2	CO	CH_4
2.42	6	8.90	0.89	0.27	[b]
12.81	10	9.75	2.12	0.93	.0023
13.31	7	6.75	2.35	1.25	.0099
15.82	5	9.60	3.58	1.09	.0061
16.75	3	8.75	2.73	1.07	.0080
18.75	12	7.00	2.92	1.15	[b]
20.66	2	8.25	3.39	1.01	.0077
22.19	16	5.25	2.87	1.56	[b]
22.85	11	6.00	3.08	0.95	.0017
24.35	4	2.73	3.16	1.77	.0143
25.84	14	5.00	3.44	1.55	.0095
27.93	15	6.50	3.65	1.06	.0053
28.34	13	5.50	3.50	1.75	.0178
28.66	18	5.58	3.85	1.65	.0084
33.93	17	5.50	4.75	1.42	.0015
34.51	1	3.00	3.72	2.04	.0025
41.00	9	3.00	4.18	1.85	.0130
47.32	8	3.25	5.02	1.72	.0133

[a] Low sample weight.
[b] No measurement obtained.

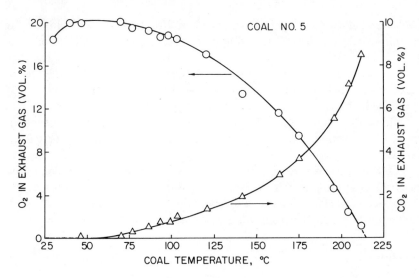

Figure 5.   *O_2 and CO_2 in exhaust gas vs. coal temperature*

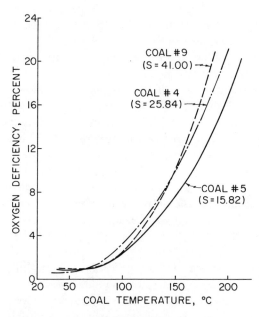

Figure 6.   *Oxygen deficiency vs. coal tem-perature*

4.56% and 10.28%, respectively, for Coal No. 18 and Coal No. 17. A minimum value of 0.01% was obtained for oxygen at 202°C (Coal No. 9). The rates of oxygen absorption and carbon dioxide production are shown graphically in Figure 5 for Coal No. 5 (S = 15.82). Only slight changes in either constituent are observed until a temperature of ~ 75° is reached. Near the end of the testing period the increase of carbon dioxide in the effluent gases is calculated to be 0.20% per degree centigrade. Correspondingly, oxygen is being absorbed at a rate of 0.26% per °C.

Oxygen deficiencies as a function of coal temperature are compared for three coal samples with considerably different S-indices in Figure 6. Differences are not dramatic; however, the curve for Coal No. 5 (with the lowest S-indice) lies somewhat below the other two.

Conclusions

The modified nonadiabatic method described for the determination of the spontaneous combustion liability of subbituminous coals can yield satisfactory results. However, a more realistic measurement of a coal's self-heating tendency might be obtained through the use of an adiabatic system. In this case, heat derived from the coal oxidation would not be transferred to the surroundings. A spontaneous combustion instrument intended for field or lab/mine applications should be less complex, yet it should be capable of producing data that is reasonably accurate for predicting the self-heating characteristics of coals.

An adiabatic method represents the most adequate technique for determining the relative tendencies of certain coals to heat spontaneously since it simulates most closely the real phenomenon. Conceivably, a field system would be similar to the adiabatic system but with appropriate modifications to hasten the oxidation process and increase the effluent gas concentrations within a reasonable test period. This could involve a more versatile system which would allow either the study of self-heating rates, similar to a method used by Guney (*10*) or which may be used for adiabatic calculations of a liability index through incorporation of a constant heat input. In the latter case, the heat might be supplied exclusively from the oxidizing air stream.

A further modification that might be incorporated into a field lab/ mine system is a more precise means of controlling the moisture content of the incoming air stream. Guney in 1972 (*11*) established the importance of moisture on the tendency of spontaneous combustion, accelerating the rate of oxidation of coal in the early stages. This influence is not only evident in the case of wet coal/wet air experiments but quite dramatic in the case of dry coal/wet air tests. Furthermore, Guney found that the heat produced owing to wetting of coal is more significant than that generated by the oxidation process. It is therefore important that instrumental features of a portable field unit incorporate means to determine, at least approximately, the difference in equilibrium humidities in coals.

Today there are several good portable gas chromatograph units on the market. Any further development of this approach for the development of a successful field method will have to consider the following criteria:

(a) simplest possible techniques for analysis without sacrificing sensitivity;

(b) compact and portable; as rugged as possible with low-power input;

(c) the capability of battery operation for at least 8 hr;

(d) simple maintenance and service;

(e) relatively fool-proof, requiring a minimum of training.

Existing portable units can incorporate both thermal conductivity and flame ionization detectors, sampling loop, switching valve, heating units for columns, etc. It appears that some modification in adsorption column configuration might be desirable since it might be somewhat difficult to maintain the first column at 0°C. This would be of little consequence in a field lab but could present problems in the field. Further work will be required to optimize these procedures for field usage.

Acknowledgment

The authors wish to acknowledge the valuable assistance provided by R. Pressey and K. Gala of the DRI Chemistry Division. This work was performed under a subcontract to D'Appolonia Consulting Engineers, Pittsburgh, Pennsylvania.

Literature Cited

1. Winmill, T. F., Graham, J. I., "Absorption of Oxygen by Coal," *Trans. Inst., Min. Eng.* (1913–1935) **46–83**.
2. *Chem. News J. Ind. Sci.* (1874) **29**, 177.
3. *Ann. Mines* (1875) **8**, 485–501.
4. Maevskaja, V. N., "Coal Classification with Reference to Spontaneous Heating," Edition Procedings NEDRA, Moscon, 1966.
5. Chamberlain, E. A. C., Hall, D. A., Thirlaway, J. T., "The Ambient Temperature Oxidation of Coal in Relation to the Early Detection of Spontaneous Heating," *Min. Eng. (London)* (1970) **121**.
6. Nand, S., Sarkar, M. K., "One-Step Analysis of a Mixture of Permanent Gases and Light Hydrocarbons by Gas Chromatography," *J. Chromatogr.* (1974) **89**, 73.

7. DiLorenzo, A., "Trace Analysis of Oxygen, Carbon Monoxide, Methane, Carbon Dioxide, Ethylene, and Ethane in Nitrogen Mixtures Using Column Selector," *J. Chromatogr. Sci.* (1970) 8(4), 224–226.
8. Smith, K. A., Dowdell, R. J., "Gas Chromatographic Analysis of the Soil Atmosphere: Automatic Analysis of Gas Samples for O_2, N_2, Ar, CO_2, N_2O, and C_1–C_4 Hydrocarbons," *J. Chromatogr. Sci.* (1973) 11, 655.
9. Kim, A. G., Douglas, L. J., "A Chromatographic Method for Analyzing Mixtures of Hydrocarbon and Inorganic Gases," *J. Chromatogr. Sci.* (1973) 11, 615.
10. Guney, M., Hodges, D. J., "An Adiabatic Apparatus Designed to Study the Self-Heating Rates of Coal," *Chem. Ind.* (1968) 1429–1433.
11. Guney, M., "Oxidation and Spontaneous Heating of Coal," *Middle East Tech. Univ. J. Pure Appl. Sci.* (1972) 5(1), 109–155.

RECEIVED October 17, 1977.

Analysis of Five U.S. Coals

Pyrolysis Gas Chromatography–Mass Spectrometry–Computer and Thermal Gravimetry–Mass Spectrometry–Computer Methods

E. J. GALLEGOS

Chevron Research Company, Richmond, CA 94802

Pyrolysis gas chromatography–mass spectrometry (py-GC–MS) and thermal gravimetry–mass spectrometry (TGA–MS), in conjunction with a dedicated computer, are used to analyze five U.S. coals under identical conditions. Data presented graphically correlate relative concentrations of various identified components with the geochemical history of the coals. The C_{27} and C_{29}–C_{30} hopanes were identified in all five coals along with several C_{15} sesquiterpanes. The concentration ratio of the 17-βH isomer to the 17-αH hopane decreases with geothermal stress experienced by the coal deposit. Other biomarker hydrocarbons, including cadalene $C_{15}H_{24}$, were identified in some of the coals. These and other details are discussed.

This paper describes the results of a detailed analysis of five U.S. coals of differing geological history using the technique of pyrolysis gas chromatography–mass spectrometry computer (py-GC–MS–C) and thermal gravimetry–mass spectrometry computer (TGA–MS–C). All data were acquired using the INCOS data system.

This paper demonstrates the fantastic power of py-GC–MS–C and TGA–MS–C for analysis of complex, nonvolatile systems such as coal. These techniques make it possible to obtain both qualitative and quantitative information on saturates, aromatics, and some hetero compounds without any prior treatment. These are on-line techniques, i.e., pyrolysis is followed by chromatographic or weight analysis which is followed by mass analysis without interruption.

0-8412-0395-4/78/33-170-013$06.00/1

Table I. Classification of Coals by Rank[a]

Btu/Lb

	I.	Anthracite	
		1. Metaanthracite	—
		2. Anthracite	—
		3. Semianthracite	—
	II.	Bituminous	
		1. Low Volatile	—
		2. Medium Volatile	—
		3. High Volatile A	—
		4. High Volatile B	14,000
		5. High Volatile C	13,000
	III.	Subbituminous	
		1. A	11,500
		2. B	10,500
		3. C	7,500
	IV.	Lignite	
		1. A	9,500
		2. B	8,300
	V.	Brown Coal	
	VI.	Peat	

(Vertical axis labels: Approx. Age ↑, Oxygen Content ↓, Heating Value ↑)

[a] Ref. *9*.

Since these are the first py-GC–MS and TGA–MS results on a series of coals, it is probably worthwhile to place the importance of coals in perspective with respect to other organic sediments as shown in Figure 1 (*1*). These world estimates are given in trillions of tons. Coal is by far the most common form of concentrated organic sediment. The United

Table II. Phanerozoic Time Scale (*3*)

Era	Period	Epoch	Beginning of Interval (Million Years)	
Cenzoic	Quaternary	Pleistocene	1.5–2	–0
		Pliocene	7	
	Tertiary	Miocene	26	
		Oligocene	37–38	
		Eocene	53–54	–50
		Paleocene	65	

Table II. Continued

Era	Period	Epoch			
Mesozoic	Cretaceous	Upper		100	−100
		Lower			
				136	
	Jurassic	Upper			−150
			162		
		Middle			
			172		
		Lower			
				190–195	
					−200
	Triassic	Upper	205		
		Middle	215		
		Lower			
				225	
	Permian	Upper	240		
					−250
		Lower			
				260	
					−300
	Carboniferous	Pennsylvanian			
				325	
		Mississippian			
				345	
					−350
Paleozoic	Devonian	Upper	359		
		Middle	370		
		Lower			
				395	
					−400
	Silurian				
				430–440	
	Ordovician	Upper	445		
		Lower			−450
				500	−500
	Cambrian	Upper	515		
		Middle			
			540		
		Lower			−550
				570	
	Precambrian				
					−600

The Encyclopedia of Geochemistry
and Environmental Sciences IVA

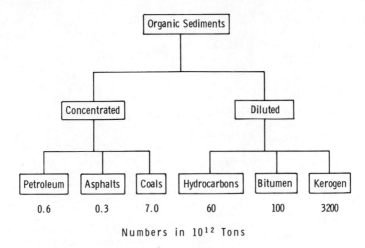

Numbers in 10^{12} Tons

Figure 1.   World estimates organic sediments distribution

States is estimated to have $\sim 1.5 \times 10^{12}$ tons of coal or ca. one-fifth of the world supply.

In the discussion of the results given in this chapter, correlations are made with respect to coal rank and geothermal stress which, in the case of the five coals studied here, coincides with geological age. Table I gives the latest ranking of coals (2) and Table II (3) gives the geological time scale.

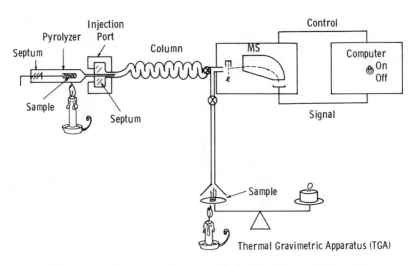

Figure 2.   System setup for pyrolysis GC–MS and TGA–MS

Experimental

The py-GC–MS work was done using a furnace pyrolyzer similar to that described previously (*4*). The pyrolyzer is coupled directly to the septum injection port of our Hewlett–Packard 7600 GC which, in turn, is coupled directly to the Nuclide 12–90–G mass spectrometer through a 200-ft, 0.02-in. ID, 5% dexil-coated capillary column. The samples were pyrolyzed at 600°C under a blanket of He carrier gas. Data were acquired at 6.5-sec intervals in the magnetic scan mode and 1-sec intervals in the multiple ion detection (MID) mode, using 20-eV ionizing voltage. Repetitive py-GC–MS–C runs using pyrolysis temperatures of 500°C–800°C did not produce significant differences in the realtive amounts of identifiable components in the molecular weight range from C_6–C_{32}.

The TGA–MS work was done using a DuPont thermal analyzer coupled directly to the Nuclide mass spectrometer through a resistance-heated capillary inlet, stainless steel tube. Both system setups are described diagramatically in Figure 2. Both air and He runs were made using a 10°C/min heating rate from ambient temperature to 700°C.

Results

The five coals analyzed are described in Table III. A map showing the location of known U.S. coal fields is shown in Figure 3. The location on this map of the five coals described here can be made by reference to Table III.

Table III.   Description of U.S. Coals Analyzed by Py-GC-MS-C and TGA-MS-C

Name	Rank	Location	Age
Loveridge (Pittsburg No. 8)	Bituminous A	West Virginia Northern	Carboniferous (Pennsylvania) $\sim 290 \times 10^6$ Yr
River King (No. 6)	Bituminous B	Illinois Southern	Carboniferous (Pennsylvania) $\sim 290 \times 10^6$ Yr
Hiawatha	Bituminous C	Utah Central	Cretaceous Upper $\sim 100 \times 10^6$ Yr
Wyodak	Subbituminous —	Wyoming Northeastern	Tertiary Paleocene 6×10^6 Yr
Noonan	Lignite —	North Dakota Northern	Tertiary Paleocene 6×10^6 Yr

TGA–MS results on the five coals are summarized in Figure 4. The low-temperature weight loss is caused by water and very light organic components. The reason for a 25°C spread between the high temperature maximum of Loveridge over the River King coal is not clear. Both show

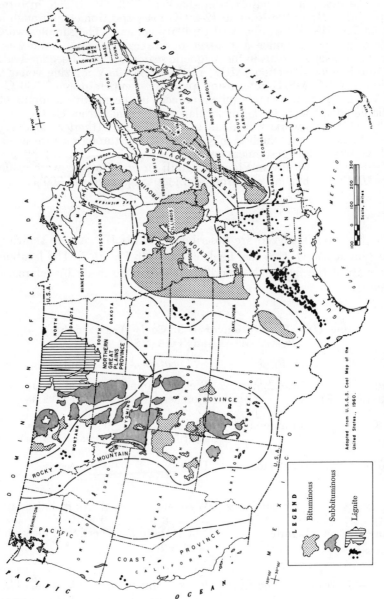

Figure 3. Bituminous and subbituminous coal and lignite fields of the conterminous United States

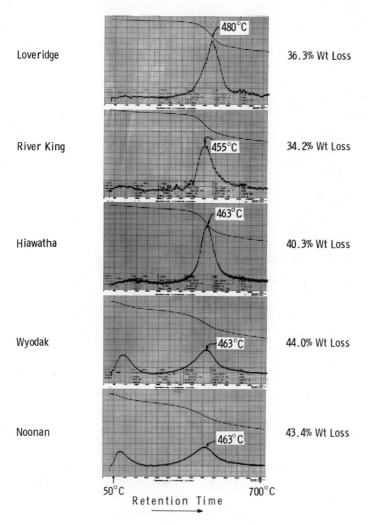

Figure 4.   TGA–MS results from five U.S. coals

Table IV.   TGA–MS—Results of Five U.S. Coals

Name	Helium Carrier Gas, Wt % Loss	Air Carrier Gas, Wt % Loss
Loveridge	36.3	93.4
River King	34.2	79.3
Hiawatha	40.3	93.3
Wyodak	44.0	93.3
Noonan	43.4	91.2

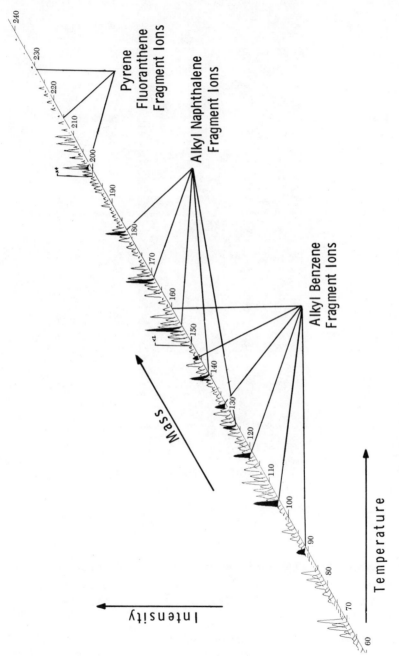

Figure 5.   Isometric display of m/e 60–260 from TGA–MS of Hiawatha coal

some weight loss between 600°C and 700°C, suggesting the decomposition of inorganic materials, perhaps carbonates, in the coal.

Similar runs were made under the same conditions using air as the carrier gas. The percent weight loss for the He and air runs are compared in Table IV. The TGA–MS–C results for Hiawatha coal summarized in Figure 5 is an example of the type of effluent to expect from this kind of

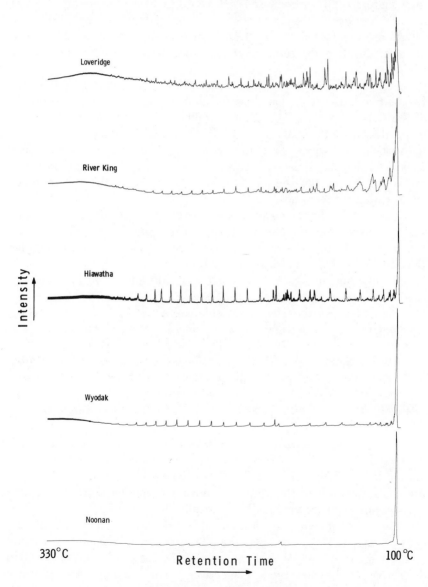

Figure 6.   Pyrograms of five U.S. coals

analysis. This is a 60-degree isometric display of the data from ca. mass 60 to $\sim$ 250. The darkened peaks are mono- to through polyaromatics as identified. The open peaks are generally caused by the saturate components present in this sample.

The py-GC–MS pyrograms for the five coals are shown in Figure 6. Data are grouped using geothermal stress as the criteria so that the most mature coal (carboniferous), Loveridge, is at the top and the least mature (tertiary), Noonan lignite, is on the bottom; all of the following figures are grouped similarly. Data shown in the following figures were all normalized to the same total ionization so that direct comparisons could be made.

Figure 7 shows the naphthalene–alkane contribution to the total for the five coals. This was done by allowing only the even-mass peaks of the C_nH_{2n+2} and C_nH_{2n-12}, i.e., paraffin and naphthalene types to be mapped in an orthogonal display. Both mass and intensity are plotted on the vertical axis, whereas retention time is plotted on the horizontal axis. There is a detectable decrease in the amount of n- and isoparaffins from the young Noonan to the old Loveridge coal. Hiawatha coal provides an exception to this trend.

The C_nH_{2n-12} shows an increase in relative concentration of naphthalenes going from the Noonan lignite to the Loveridge coal. Cadalene was identified only in the Noonan, Wyodak, and Hiawatha samples. Bendoritas (5) first identified cadalene in three crudes from the Jackson Sands formation in Texas. He suggested naturally occurring cadinene as a possible precursor.

Figure 8 shows the even-mass peak map of the C_nH_{2n-6} and C_nH_{2n-20} components. These correspond to the alkylbenzene, phenanthrene, and anthracene monocyclics and their alkyl derivative types, respectively.

All five samples show benzene, toluene, the xylenes, and a number of higher molecular weight alkylbenzenes. In addition, all show a homologous series of n-alkylbenzenes from ca. C_9–C_{27}. These show relatively intense m/e 92 (shown) and 91 (not shown) fragment ions, suggesting a homologous series of n-alkylbenzenes.

The n-alkylbenzenes decrease in concentration going from the young to the older coals. The reverse is true of the monocyclic phenanthrene–anthracenes. Though not confirmed by coinjection, copaene $C_{15}H_{24}$ is present apparently in the younger coals. The mass spectra of copaene and the unknown are practically identical.

Figure 9 shows five orthogonal even-mass peak maps of the C_nH_{2n-8} and C_nH_{2n-22} components in the five coals. These correspond to the benzomonocyclics and the pyrene, fluoranthene, and their alkyl derivative hydrocarbon types, respectively. There is an obvious increase in concentration of the alkyl pyrenes–fluoranthenes with the age of the

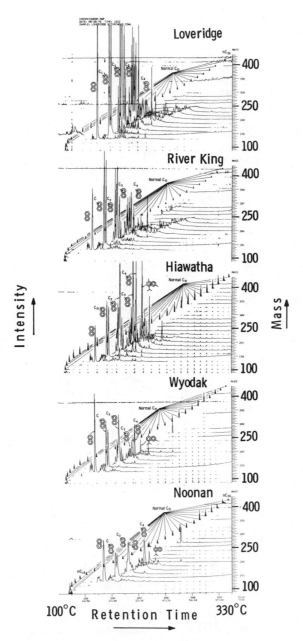

*Figure 7. C_nH_{2n+2}, C_nH_{2n-12} naphthalene, alkane
maps of five U.S. coals*

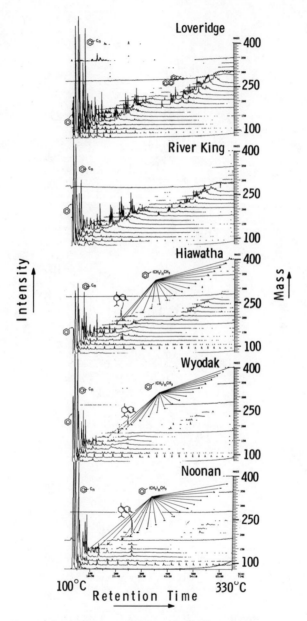

Figure 8. C_nH_{2n-6}, C_nH_{2n-20} alkylbenzene, monocyclic phenanthrene, anthracene maps of five U.S.
coals

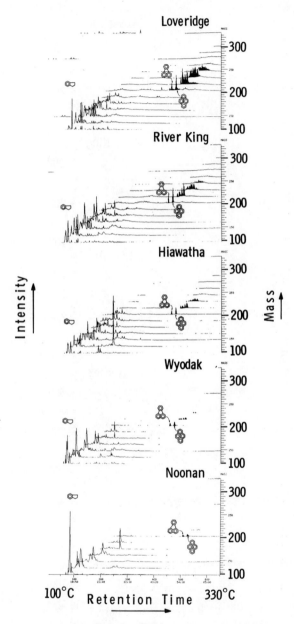

Figure 9. C_nH_{2n-8}, C_nH_{2n-22} benzomonocyclic,
pyrene–fluoranthene maps of five U.S. coals

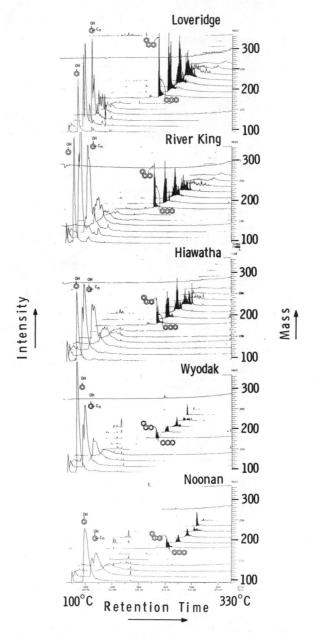

*Figure 10. $C_nH_{2n-6}O$, C_nH_{2n-18} phenol, anthracene–
phenanthrene maps of five U.S. coals*

coals. The benzomonocyclics' total concentration remains reasonably constant with age.

Figure 10 shows the orthogonal even mass maps of the $C_nH_{2n-6}O$ and C_nH_{2n-18} components that correspond to phenol–alkyl phenols and the anthracene–phenanthrenes and alkyl derivatives, respectively. There is a clear increase in relative concentration of these polyaromatics with the

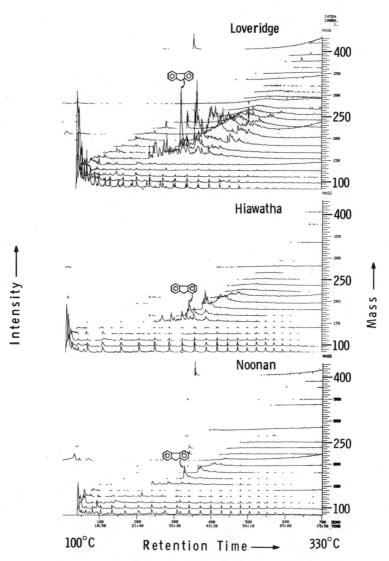

Figure 11. C_nH_{2n-2}, C_nH_{2n-16} *dialkene, cyclic alkene, fluorenes maps of three U.S. coals*

age of the coal. The ratio of phenol to the alkylphenols decreases rapidly with the age of the coal. Although it may be difficult to assess from the map, total phenols also decrease somewhat with the age of the coal.

Figure 11 shows orthogonal maps of only the Noonan, Hiawatha, and Loveridge coals. This map shows the C_nH_{2n-2} components, including the dialkenes or monocyclic alkene-type components which appear to

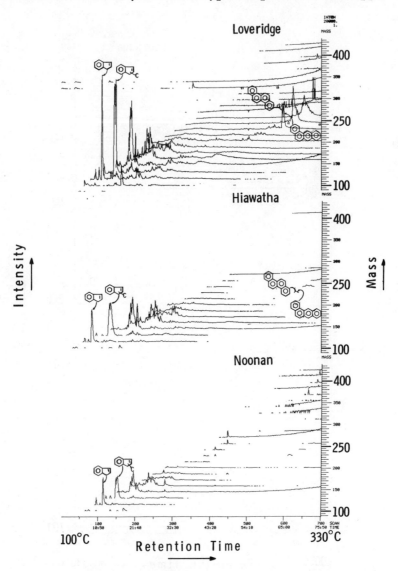

Figure 12. C_nH_{2n-10}, C_nH_{2n-24} indene-tetraaromatic maps of three U.S. coals

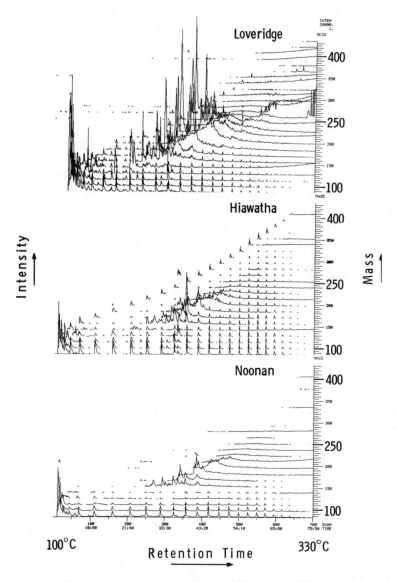

Figure 13. C_nH_{2n-0}, C_nH_{2n-14} *alkene, cyclic maps of three U.S. coals*

increase with the age of the coal. Only the even-mass fragment ions for these component types appear in this map; the molecular ions are too low in intensity to be seen.

Also shown on this map are the C_nH_{2n-16} components which, as it turns out, are mainly caused by the fluorenes. The fluorenes increase in relative concentration with the age of the coal.

Figure 12 shows the even-mass ion map of the C_nH_{2n-10} which corresponds in this case to indenes and the C_nH_{2n-24} components which are the tetraaromatics. Only the oldest coal, Loveridge, shows significant amounts of these tetraaromatics. The chrysene and benzanthracene shown in this figure were not confirmed by coinjection.

Finally, Figure 13 is an attempt to show the C_nH_{2n-0} components that correspond to the alkenes or monocyclics and the C_nH_{2n-14} components that correspond to acenaphthene and alkyl derivative-type hydrocarbons. The C_nH_{2n-0} components, either monocyclics or alkenes, appear to increase in concentration with the age of the coal. Both Hiawatha and Loveridge coals show the molecular ions of the alkenes. They appear as a homologous series of doublets. One appears just before the n-alkane and one just after. The C_nH_{2n-14} components are wiped out essentially because of interference with fragment ions (filled peaks) from the naphthalenes which are of relatively high concentration in these samples.

The average of peaks down to the lower masses are caused by even-mass fragments from the homologous series of alkenes.

Figure 14 shows the C_{27} and the C_{29}–C_{32} terpane m/e 191 mass chromatographic profiles for the five coals. Their mass spectra match those of the hopane series found abundant in Messel shale (6). The C_{27} and C_{29}–C_{30} hopanes were confirmed by coinjection with the authentic hopanes. Note the decrease in relative concentration with the maturity of the 17-βH C_{27}, C_{29}, and C_{30} hopanes (7). This is further confirmation of the thought that the 17-βH hopanes are thermodynamically less stable than the 17-αH, 21-βH series. The ratios of the 17-βH series to the 17-αH series may be reliable indicators of the maturity of a deposit. This may be complicated by the chance that the relative intensities within a series may carry some information about environmental conditions at the time of formation of a deposit.

Figure 15 shows isometric maps of Loveridge and Hiawatha coals and Noonan lignite. Shown on these maps are the mass chromatograms of m/e 191 and 206, and the molecular ions' masses corresponding to the C_{27}, C_{28}, C_{29}, C_{30}, and C_{31} pentacyclic triterpanes.

The kinds of aromatics that contribute to m/e 191 and 206 are the phenanthrene–anthracenes and their alkyl derivatives.

There are two observations to be made here. The relative concentration of the sesquiterpanes to the pentacyclic diterpanes decreases with the maturity of the coal. The relative concentration of total terpane to aromatics also decreases with the maturity of the coals.

Figure 16 is a 30-degree isometric cap of all the data acquired in a py-GC–MS run of a Hiawatha coal sample. This figure serves to give an

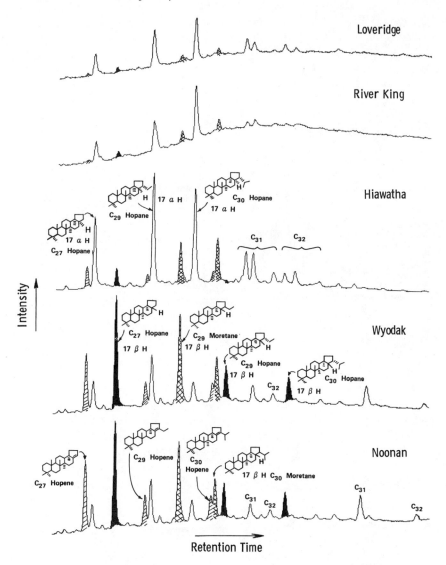

*Figure 14.   The m/e 191 mass chromatograms
showing triterpane profiles of five U.S. coals*

overview of the relative importance of various components in the total
sample. The darkened peaks are caused by the homologous series of
alkanes and alkenes; the phenols are responsible for the cross-hatched
peaks; and the open peaks mainly are caused by the aromatics. Cadalene
is identified; also the hopanes are identified from their *m/e* profile.

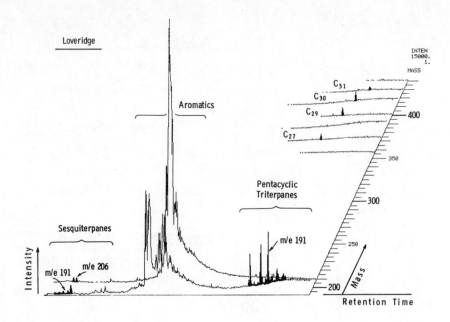

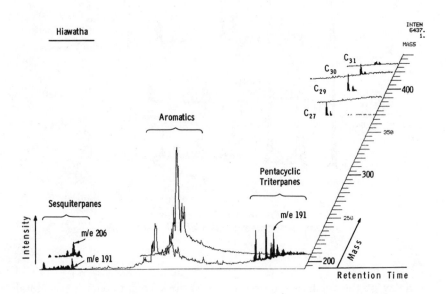

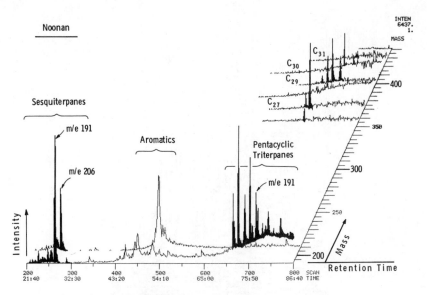

Figure 15.   Isometric map of three coals showing the relative importance of triterpanes to sesquiterpanes to aromatics

Discussion

Several points revealed as a result of this study are listed as follows:

1. Normal and branched alkanes, benzene–alkylbenzenes, copaene, cadalene, phenols, sesquiterpanes, and pentacyclic triterpanes decrease in relative concentration with the maturity of the coal.

2. Naphthalenes, monocyclic phenanthrene–anthracenes, pyrenes, fluoranthrenes, phenanthrenes, anthracenes, fluorenes, indenes, alkenes, and polyaromatics generally increase with the maturity of the coal.

3. The benzomonocyclic saturates' total concentration appears to hold constant, although individual components within these types change concentration relative to each other over the geological time and stress experienced by the deposit.

All of the points listed above support a traditional and accepted concept of the maturation of coal, i.e., decrease in oxygen concentration and graphitization of the organic remains with maturity.

4. The 17-βH, 21-βH hopane concentrations relative to the 17-αH isomer appear to be very sensitive indicators of the rate of maturation of sediments. The total terpane concentration changes from ~ 1.1% in Noonan lignite to about 0.4% in the Loveridge coal.

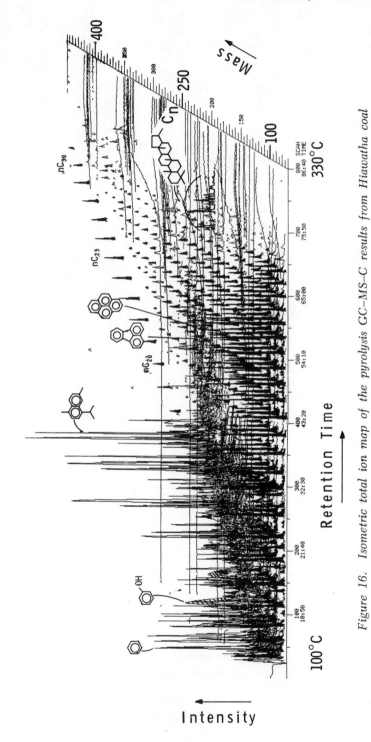

Figure 16. Isometric total ion map of the pyrolysis GC–MS–C results from Hiawatha coal

5. Cadalene and copaene found only in the three youngest coals suggests that they may be used as maturity or enivornmental monitoring components.

Terpanes and steranes, as well as their partially aromatized counterparts, cadalene, and copaene released from nonvolatile insolubles in coal and kerogen from shales by pyrolysis are contamination-free biomarkers that carry information concerning their geochemical history.

Although coal and kerogen in shale do not migrate, continents do. This could lead to useful correlation studies similar to those done presently on petroleum reserves and source rocks. Presently, biological marker analysis by GC–MS, together with isotope analysis and geological considerations, are used to characterize known oil fields. Using this and correlation data from source rocks and reservoirs facilitates prediction of possible location, age, and depth of new reservoirs.

This same approach could be used both to validate further the presently entrenched continental drift theory and predict from coal and shale analysis of an area of one continent the type of organic deposits which may be present on another fitted but as yet unexplored continent (8).

A last but important aspect of this work is the demonstrated ability to pull out specific, qualitative, and quantitative information from a complex material such as coal using py-GC–MS and TGA–MS.

2097493

Acknowledgments

The author would like to thank L. P. Lindeman and A. L. McClellan for helpful suggestions during the preparation of this paper. Thanks also to W. K. Seifert and J. M. Moldowan for the authentic hopanes used in coinjection.

Literature Cited

1. Yen, T. F., "Developments in Petroleum Sciences 5, Oil Shale," T. F. Yen, G. V. Chilingarian, Eds., pp. 129–148, Elsevier, Amsterdam, 1976.
2. American Society for Testing and Materials, Annual Book of ASTM Standards, Designation: D 388–66, Pt. 19, pp. 54–58, 1972.
3. Goldich, S. G., "The Encyclopedia of Geochemistry and Environmental Sciences IVA," Fairbridge, R. W., Ed., pp. 453–455, Van Nostrand Reinholt, 1972.
4. Gallegos, E. J., *Anal. Chem.* (1975) **47**, 1524–1528.
5. Bendoritas, J. G., "Advances in Organic Geochemistry," B. Tisot, S. Bienner, Eds., pp. 209–224, Pergamon, London, 1973.
6. Kimble, B. J., Maxwell, J. R., Philp, R. P., Eglinton, G., Arpino, P., Albrecht, P., Ourisson, G., *Geochem. Cosmochim. Acta* (1974) **38**, 1165–1181.

7. Ensminger, A., Van Dorsselaer, A., Spyckerelle, C., Albrecht, P., Ourisson, G., "Advances in Organic Geochemistry," B. Tisot, S. Bienner, Eds., pp. 245–258, Pergamon, London, 1973.
8. Hurley, P. M., *Scientific American* (April, 1968) 53–64.
9. American Society for Testing and Materials, 1972 Annual Book of ASTM Standards, Designation: D–388–66, pp. 54–58.

RECEIVED August 5, 1977.

An ^{1}H and ^{13}C NMR Study of the Organic Constituents in Different Solvent-Refined Coals as a Function of the Feed Coal

D. L. WOOTON, W. M. COLEMAN, T. E. GLASS,
H. C. DORN, and L. T. TAYLOR

Department of Chemistry, Virginia Polytechnic Institute & State University,
Blacksburg, VA 24061

The tetrahydrofuran (THF)-soluble portion ($> 90\%$) of a series of solvent-refined coals derived from different feed coals has been preparatively separated into sized fractions via gel permeation chromatography. Each preparative separation regardless of the coal was divided into four fractions based on a constant elution volume. Weight distributions and average molecular weights have been measured for each of the four fractions. Organic constituents were analyzed for the major fractions via quantitative ^{1}H and ^{13}C Fourier transform nuclear magnetic resonance (NMR) techniques in order to determine various molecular parameters such as H/C ratios for aliphatic and aromatic components, weight percent carbon and hydrogen, and percent aromaticity. Inferences on the chemical structures of the organic constituents present in the same fraction from different coals are described. Sample solubility and homogeneity limited the precision and accuracy of the high-resolution NMR approach.

The solvent-refined coal (SRC) process is one of several processes currently under consideration for converting coal to a relatively ash-free, low-sulfur fuel (*1*). Chemical characterization of the heterogeneous semisolid SRC product obtained from this process should provide a better understanding of the SRC process and perhaps new insight regarding coal processing in general. An important variable in the SRC process is the type of feed coal used. In this study we focus attention

0-8412-0395-4/78/33-170-037$05.00/1

on the nature and molecular composition of various sized fractions of different SRC's which differ in the geographical origin of their feed coal.

The large number and molecular complexity of the components in a single SRC solid product (2) represent a formidable task for any type of chemical analysis. This is even more acute for the present study which involves monitoring chemical changes as a function of several coal processing variables (e.g., temperature, solvent, feed coal, etc.). The chromatographic methods used normally in this area of study involve initial chromatographic separation usually into fractions based on either size or type (e.g., nonpolar, polar). The latter approach (3) has the advantage of potential separation of the coal product mixture into fractions consisting of several chemical classes, such as aliphatics, aromatics, acids, bases, etc. This approach, however, is usually nonpreparative and it suffers from the fact that less than quantitative recovery from the chromatography column is achieved generally. The alternate approach of separation according to effective molecular size was used in the present study and involved separation of the SRC solid product via high performance gel permeation chromatography (GPC). This approach allows routine preparative separations with close-to-quantitative recovery of the material injected on the gel permeation column. The details of the separation and its application to one of the coals in this study have been published previously (4).

It should be emphasized that this chromatographic method provides only sized fractions. Thus oxygen, nitrogen, and sulfur as well as other functional groups usually are distributed randomly in the fraction collected. Chemical characterization of these sized fractions in terms of average molecular parameters is less than ideal, but it should provide insight regarding prominent chemical changes as a function of the feed coal used in the SRC process.

With a chromatographic technique capable of routinely yielding preparative fractions, quantitative ^1H and ^{13}C FT NMR was the major spectroscopic tool used for chemical characterization. The established utility of ^1H and ^{13}C NMR for characterization of coal products is documented well. Unfortunately, high-resolution ^{13}C FT NMR is not quantitative normally under operating conditions used typically. (It should be noted that quantitative ^1H FT NMR measurements also are not obtained routinely. The problem of variable spin lattice relaxation times (T_1's) is present also in ^1H FT NMR. In addition, the greater signal intensity of ^1H NMR in comparison with ^{13}C FT NMR poses an additional potential problem of detector linearity in the ^1H FT NMR receiver.) For ^{13}C FT NMR, variable spin lattice relaxation times (T_1's) and nuclear Overhauser effects (a result of pseudo random ^1H noise decoupling) usually cause wide variations in signal intensity from line to line within a given spectrum. To overcome this problem, addition of a paramagnetic relaxation reagent to the NMR sample along with a gated ^1H decoupling sequence provides a method for obtaining quantitative ^{13}C NMR data (5). This technique in conjunction with the addition of a standard to the NMR sample provides a method for obtaining a number of important molecular parameters such as total, aromatic, and aliphatic hydrogen to carbon ratios, $(H/C)_{tot}$, $(H/C)_{ar}$, and $(H/C)_{al}$, respectively. In addition, if weighed quantities of the SRC fraction as well as the reference standard are used, weight percent carbon and hydrogen values can be determined without recourse to elemental combustion data. This is an important consideration for studies of complex mixtures encountered usually in studies of coal products since the solution NMR measurements may not reflect the bulk results obtained by elemental combustion. That is, the ^1H and ^{13}C quantitative NMR approach complements the carbon and hydrogen elemental combustion data by providing a means of comparing the weight percent hydrogen and/or carbon observable in the NMR experiment with the total values from elemental combustion.

^1H and ^{13}C NMR, elemental combustion, and average molecular weight data (via vapor-phase osmometry) provide sufficient information for assessing several average molecular parameters of the sized fractions (e.g., the number of aromatic carbons (C_{ar}) and hydrogens (H_{ar}) per average molecule). Although characterization of the sized fractions in terms of average molecular parameters provides insight regarding average molecular structure(s), these structures should be viewed with due caution, and are not necessarily representative of the individual molecules actually present in a given fraction. However, in the present study it was anticipated that comparison of the average molecular parameters (e.g., C_{ar}, C_{al}, etc.) as a function of the sized fractions derived from different SRC samples could provide a convenient means of monitoring significant chemical changes in coal processing variables.

Results and Discussion

The five SRC samples examined in this study were obtained from a pilot plant operating at Wilsonville, Alabama. We have reported previously extensive data for the SRC derived from Pittsburgh #8 feed coal and this sample naturally will serve as the reference SRC sample (6). The other SRC samples were derived from Western Kentucky #9 and #14 coal, Illinois #6 coal, high-sulfur Illinois coal, Monterey, and a western coal, Amax. We hoped that the coals from different regions would yield significant chemical differences in their respective SRC products. Following this argument the Amax SRC sample might be

Table I. Pilot Plant Processing Conditions[a]

Feed Coal	Temperature	Conversion	Reaction Time (min)	H_2 Pressure (psi)	% H_2 Consumption
Pittsburgh	855°F	90%	26.4	1700	2.7
Amax	853°F	79%	20.5	2400	3.3
Monterey	837°F	95%	26.5	2400	3.8
Illinois	824°F	92%	20.7	1700	3.5
Western Kentucky	853°F	90%	15.2	1700	2.1

[a] Data supplied by Southern Services Inc., Montgomery, Alabama.

expected to yield a significantly different SRC product when compared with an eastern SRC (e.g., Pittsburgh #8 SRC). However, it should be noted that some of the differences between SRC solid products may reflect different pilot plant processing conditions. Pertinent reaction parameters in effect when these SRC samples were drawn are listed in Table I.

Preparative separation of the THF-soluble portion of each SRC was accomplished using high performance GPC. A typical chromatogram for the Amax sample using a refractive index detector is presented in Figure 1. Preliminary studies indicate that a reasonable choice for the column

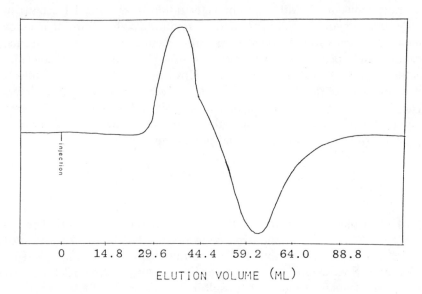

ELUTION VOLUME (ML)

Figure 1. Glass column (see Ref. 4 for dimensions) packed with Bio-Beads SX–4 (200–400 mesh); THF flow rate, 2.8 ml/min; pressure 30 psi; detection, refractive index; sample, 0.44 g of THF-soluble Amax SRC dissolved in 15 mL of THF

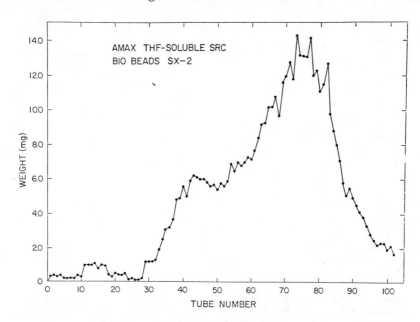

Figure 2.   *Chromatographic fraction weight profile for Amax THF-soluble SRC using Bio-Beads SX–2. See Figure 1 for experimental parameters.*

packing material for this study is Bio-Beads SX–4 (4% cross-linking). In support of this choice, weight profiles obtained for the same Amax SRC sample are presented in Figures 2, 3, and 4 with the column packing material Bio-Beads SX–2, SX–4, and SX–8, respectively. For the SX–8 (8% cross-linking) packing with relatively low exclusion limits (Figure 4), the major portion of the SRC is eluted rather early in the chromatogram. Whereas, with the SX– (2% cross-linking) packing most of the molecules appear at much longer retention volumes (Figure 2). In Figure 3, the weight profile for the Amax SRC sample utilizing Bio-Beads SX–4 is indicated with the arbitrary elution volume cuts for Fractions 10, 20, 30, and 40. The same arbitrary elution volumes were used for each of the SRC samples (i.e., Pittsburgh #8, Monterey, etc.) in each preparative separation. Cursory examination of the weight profiles and/

Table II.   Solubility and Fractional Weight Distributions

THF-Soluble SRC	% Soluble	Recovered Material Distribution				% Recovery
		10 (%)	20 (%)	30 (%)	40 (%)	
Amax	91.8	27.4	14.3	57.7	0.6	94.3
Monterey	100	30.5	20.7	47.4	1.4	100
Pittsburgh	97.1	30.8	18.8	47.4	3.0	97.7
Illinois	91.2	43.0	21.9	33.3	1.8	98.9
Western Kentucky	88.4	32.1	19.3	43.3	5.3	97.5

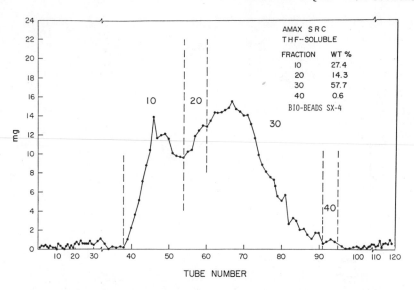

Figure 3. Chromatographic fraction weight profile for Amax THF-insoluble SRC using Bio-Beads SX–4. See Figure 1 for experimental parameters.

or the chromatograms for each of the SRC samples reveals that they are very similar when the same chromatographic conditions are used. The weight distribution data for each fraction (#10–40) as a function of the different SRC sample are presented in Table II.

The Amax sample contains a significantly lower amount of high molecular weight material (Fractions 10 and 20 ≃ 42%), when com-

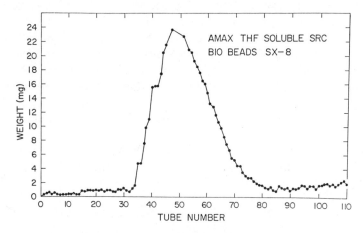

Figure 4. Chromatographic fraction weight profile for Amax THF-soluble SRC using Bio-Beads SX–8. See Figure 1 for experimental parameters.

pared with the Illinois sample (~ 59%). However, the weight percent in each fraction for the other SRC samples remains relatively constant (Fractions 10 and 20 ≃ 50%). In addition, Table II contains solubility and chromatography percent sample recovery data for each SRC sample indicating the high THF solubility and sample recovery obtainable with this chromatographic approach. The solubility of each SRC was determined by placing approximately 1.0 g of SRC in 250 mL of THF and stirring for 2 hr. The solution at room temperature was filtered through a Millipore filtering apparatus equipped with a LWSP filter. Both soluble and insoluble portions were dried overnight in vacuo (10^{-6} torr) after removal of THF prior to weighing.

Average molecular weight data were obtained for Fraction 30 for each SRC sample (Table III) using vapor-phase osmometry measurement in THF. A reasonable expectation based on the chromatographic ap-

Table III.   Molecular Weight Data[a] for Solvent-Refined Coals

	Feed Coal				
SRC Sample	Pitts-burgh	Amax	Mon-terey	Illi-nois	West-ern Ken-tucky
THF-soluble (THF) [b]	626	641	672	574	691
CHCl$_3$-soluble (THF) [b]	541	535	535	600	542
CHCl$_3$-soluble (CHCl$_3$) [b]	684	692	672	593	737
Fraction #30 (THF) [b]	388	326	547	374	313

[a] Vapor-phase osmometry; mol wt values are those obtained by extrapolating to infinite dilution: concentration range for experimental data = 0.6–2.0 g/100 mL.
[b] Solvent in which mol wt measurements were performed.

proach would suggest relatively constant average molecular weight values for a given fraction independent of the nature of the SRC sample. This is generally true; however, an exception does occur as indicated by the relatively high value (547) obtained for the Monterey sample.

Ideally, for studies of this type the same solvent used in the chromatographic separation should be the solvent of choice for obtaining the ^{1}H and ^{13}C NMR data. Unfortunately, THF is a poor choice as a NMR solvent because of spectral solvent interferences in the ^{1}H and ^{13}C spectral regions of interest. The solvent used in the ^{1}H and ^{13}C NMR measurements was chloroform-d which is not nearly as efficient as THF in dissolving the original SRC samples. This decrease in solubility is crucial regarding discussions of the NMR data to follow, but it is also apparent from the molecular weight data presented in Table III. The average molecular weight data in Table III are for the original CHCl$_3$-

and THF-extracted SRC samples measured in either THF and/or $CHCl_3$. The consistently lower average molecular weights (except Illinois SRC) obtained for the $CHCl_3$-soluble SRC samples in THF solvent vs. the THF-soluble SRC samples in the same solvent apparently reflect the poorer ability of $CHCl_3$ to dissolve the higher molecular weight components in these SRC samples. As support for this postulate, the higher molecular weight components exhibit much lower solubility in $CHCl_3$ than THF (e.g., Fraction #10 samples have only limited solubility in $CHCl_3$). An explanation for the failure of Illinois SRC to behave in this manner is not apparent. The molecular weight data in Table III also indicate significant differences for the same SRC when measured in $CHCl_3$ or THF. $CHCl_3$-soluble SRC in $CHCl_3$ yields molecular weight values ≈ 150 units higher (except Illinois) than $CHCl_3$-soluble SRC in THF. The poorer donor ability of $CHCl_3$ over THF which could result in a higher degree of association between coal molecules may account for this observation. The abnormal behavior of Illinois in this regard should be noted again.

Analytical data obtained from elemental combustion analysis for Fraction 30 of the five SRC samples are presented in Table IV. The relatively constant molecular size of each Fraction 30 sample with 22–28 carbon atoms per average molecule is noteworthy. The only exception which has been noted previously in terms of the mol wt data is the Monterey Fraction 30 sample with 38 carbon atoms per average molecule. The total hydrogen to carbon ratio $(H/C)_{tot}$ is lower for each Fraction 30 sample than $(H/C)_{tot}$ obtained for the non-chromatographed THF-soluble SRC sample. Perhaps one of the more significant trends is the consistently lower $(H/C)_{tot}$ values for the western SRC Fraction 30 (Amax, 0.77; Monterey, 0.79) relative to the other Fraction 30 samples (0.86–0.90).

Table IV.   Analytical Data for

Ultimate Analyses

Sample	% C	% H	% N	% S
W. Kentucky	83.7	6.3	1.7	1.6
Pittsburgh[a]	86.6	6.1	1.8	0.9
Illinois	88.0	6.3	1.7	1.4
Monterey	83.8	5.5	1.6	1.5
Amax	87.6	5.6	1.2	0.8

[a] Values measured in this study for the Pittsburgh #8 Fraction 30 sample are in general agreement with values previously reported (6).

[b] $(H/C)_{tot}$ values in parenthesis are for the non-chromatographed THF-soluble portion of each SRC sample.

Representative ^1H and ^{13}C spectra are presented in Figures 5, 6, 7, and 8 for the Amax and Monterey samples. The quantitative reference (peak at 0.0 ppm) in each spectrum is hexamethyldisiloxane ($(CH_3)_3$-Si-O-Si$(CH_3)_3$). A paramagnetic relaxation reagent (the paramagnetic relaxation reagents used were either tris(acetylacetonato)iron(III), Fe(acac)$_3$, or tris(acetylacetonato)chromium(III), Cr(acac)$_3$ at concentrations of $2-6 \times 10^{-3}M$). was added to decrease spin lattice relaxation times (T_1's) and suppress nuclear Overhauser effects (5). In addition, ^1H gated decoupling was used for the ^{13}C NMR measurements where the ^1H decoupler was on only during the time interval that the ^{13}C magnetization was monitored (t) and was off for a longer time (T). Typically, the values for t and T were 0.65 and 4.35 sec, respectively. In most cases the value $T = 4.35$ seconds was found to be sufficiently long so that the ^{13}C integrals did not vary with longer T values indicating complete recovery of the ^{13}C magnetization.

Superficially, the spectra for all Fraction 30 samples are very similar. One noticeable difference is the Monterey ^{13}C spectrum (Figure 6) which contains a relatively greater number of signals in the aromatic region below 130 ppm. This indicates the greater importance of larger condensed aromatic rings and/or heterocyclic aromatic ring carbons present in this sample. This is consistent with the larger average molecular formula and greater hetero atom content indicated in Table IV.

As previously indicated, quantitative ^1H and ^{13}C NMR provide methods of dissecting the average molecular formula data into average molecular parameters such as H_{al}, H_{ar}, C_{al}, and C_{ar}. Furthermore, the quantitative approach (5) provides a method for obtaining H/C ratios along with weight percent carbon and hydrogen data which can be compared with the elemental combustion data. We have utilized this latter approach for model studies (5), the Pittsburgh #8 SRC sample (6), and

SRC Fraction 30 Samples

$(H/C)_{tot}$	Average Molecular Formulas[c]			
0.90 (0.97)[b]	$C_{21.8}$	$H_{19.6}$	$N_{0.4}$	$S_{0.1}$	$O_{1.4}$
0.85 (0.89)[b]	$C_{28.0}$	$H_{23.7}$	$N_{0.5}$	$S_{0.1}$	$O_{1.1}$
0.86 (0.94)[b]	$C_{27.4}$	$H_{23.5}$	$N_{0.5}$	$S_{0.2}$	$O_{0.6}$
0.79 (0.92)[b]	$C_{38.2}$	$H_{29.1}$	$N_{0.6}$	$S_{0.3}$	$O_{1.3}$
0.77 (0.88)[b]	$C_{23.8}$	$H_{18.3}$	$N_{0.3}$	$S_{0.1}$	$O_{1.0}$

[c] Average molecular formulas obtained from elemental combustion and av mol wt data (vapor-phase osmometry) with the oxygen values determined by difference.

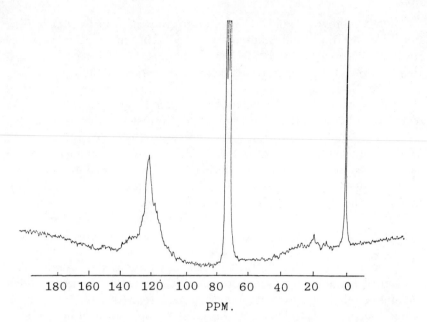

Figure 5. ^{13}C *FT NMR spectrum of Amax Fraction 30 in d-chloroform with hexamethyldisiloxane reference*

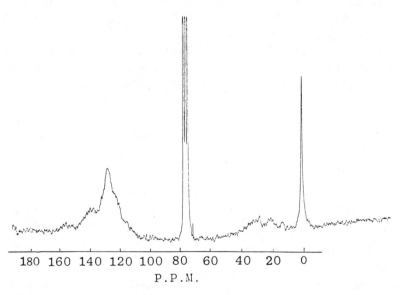

Figure 6. ^{13}C *FT NMR spectrum of Monterey Fraction 30 in d-chloroform with hexamethyldisiloxane reference*

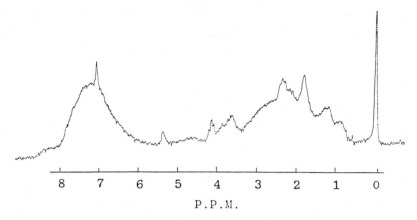

P.P.M.

Figure 7. *¹H FT NMR spectrum of Amax Fraction 30 in d-chloroform with hexamethyldisiloxane reference*

other fuels derived from coal and shale (7). The weight percent carbon and hydrogen data obtained directly from NMR spectra provide a means of checking whether all the carbon or hydrogen is observed in the NMR experiment relative to the bulk elemental combustion analysis. This is an important consideration since the NMR measurements are normally solution measurements. The weight percent carbon values obtained from direct ¹³C NMR measurements on 5–10 different samples for fraction 30 samples of Pittsburgh, Western Kentucky, and Illinois were 80.3, 87.3, and 103.7, respectively with standard deviations rangin from 5–15%. These values are in fair agreement with the elemental combustion data presented in Table IV recognizing that this approach is mainly useful in

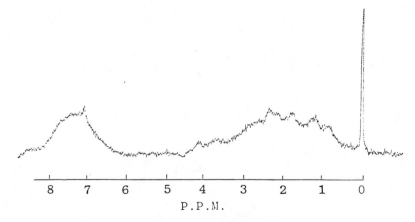

P.P.M.

Figure 8. *¹H FT NMR spectrum of Montery Fraction 30 in d-chloroform with hexamethyldisiloxane reference*

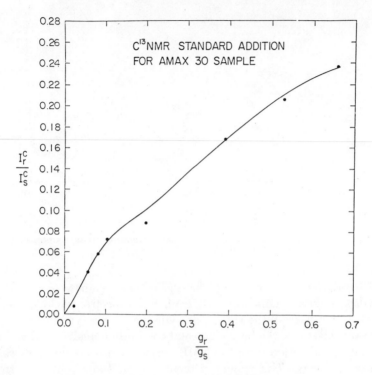

Figure 9. ¹³*C integrated intensity ratio of Amax Fraction 30 and hexamethyl-disiloxane vs. the weight ratio of hexamethyldisiloxane to Amax Fraction 30*

confirming that the majority of the carbon atoms are observed in the NMR experiment. It should be noted, however, that the standard deviations are much larger than observed previously in related studies (5, 6, 7). Furthermore, the standard deviations were even greater for Amax and Monterey Fraction 30 samples ($\sim \pm 25\%$). The problem for the Amax 30 sample is indicated in Figure 9 which is a plot of the ¹³C integrated intensity ratio of the sample and the reference (hexamethyldisiloxane) ($I[C_r/C_s]$) vs. the weight ratio of the reference to sample (g_r/g_s). Ideally, this plot should be linear with an intercept at zero when the slope provides the weight percent carbon from the equations below:

$$\frac{\text{slope}}{(^{13}\text{C NMR})} = \left(\frac{\#C_r}{\text{mol wt}_r}\right)\left(\frac{\text{mol wt}_s}{\#C_s}\right) \qquad (1)$$

$$\text{wt-\% C} = \frac{\#C_s}{\text{mol wt}_s} \times 1200 \qquad (2)$$

The plot in Figure 9 was obtained by adding four incremental weighed amounts of reference to the Amax SRC 30 sample for the four points with the smallest (g_r/g_s) ratio. This procedure was repeated for the four points with the higher (g_r/g_s) ratio utilizing the identical Amax SRC sample. As indicated in Figure 9 the slope changes dramatically at $(g_r/g_s) \simeq 0.2$. The slope for the separate lower four points and the upper four points indicate percent carbon values of 67.3 ± 4.1 and 111.3 ± 10.8, respectively. After examination of the quantitative ^{13}C NMR procedure for possible sources of error in terms of the NMR instrumentation and applying the technique in several model compound studies, we conclude that this problem is probably caused by an inhomogeneous solution (e.g., colloidal suspension) from the NMR point of view. (The relatively low signal-to-noise ratios for the ^{13}C spectra and corresponding baseline roll is one source of error in the ^{13}C NMR integrations in the present study. However, changing the curvature and slope of the baseline utilizing the FT computer program and comparisons with other data $(5, 6, 7)$ suggests that this source of error is no greater than ± 5% which is on the order of the standard deviation quoted.) Two sources of non-homogeneity can be envisioned. These are partial precipitation of the coal during the NMR measurement and incomplete solubility of the reference in the solvent. It should be noted that the solutions used in the NMR experiment were filtered. The incompatibility of the particular solvent and reference combination used cannot be the entire problem since the aromatic to aliphatic carbon ratio (C_{ar}/C_{al}) for the sample changes significantly between the lower and upper (g_r/g_s) regions, 3.83 and 3.12, respectively. The general problem described above is reflected also in the ^1H NMR data. This problem was encountered also with the Monterey SRC 30 sample and could be present to a lesser degree for the other three SRC fractions mentioned previously.

The general problem of solution homogeneity represents a major area of concern in high-resolution solution NMR studies of complex hydrocarbon mixtures encountered in coal products. Although alternate NMR approaches could be used or developed (e.g., solid-state NMR, high-temperature NMR, ideal NMR solvents, etc.), the quantitative NMR approach used in the present study can distinguish this problem when complementary elemental combustion data is available.

The average molecular parameters obtained from the ^1H and ^{13}C NMR data are summarized in Table V. The rather low $(H/C)_{ar}$ ratios (0.38–0.58) indicate the presence of aromatic rings which are either highly condensed and/or substituted with alkyl substituents. For most samples, 1–5 aromatic rings per average molecule could be inferred from the data. However, the Monterey SRC Fraction 30 exhibits the lowest aromatic ratio suggesting an even greater number of condensed aromatic

Table V. Average Molecular Parameters for SRC Fraction 30 Samples

Sample	$H_{al}{}^a$	$H_{ar}{}^a$	$C_{al}{}^a$	$C_{ar}{}^a$	$(H/C)_{ar}{}^a$	$(H/C)_{al}{}^a$	$f_a{}^b$
Western Kentucky	12.7	6.9	6.3	15.5	0.44	2.01	(0.71)
Pittsburgh	15.4	8.3	10.1	17.9	0.46	1.52	(0.64)
					(0.49)	(1.66)	
Illinois	12.7	10.8	8.8	18.6	0.58	1.44	(0.68)
Monterey	17.8	11.3	8.8	29.4	0.38	2.02	(0.77)
Amax	8.9	9.4	5.4	18.4	0.51	1.65	(0.77)
					(0.52)	(1.67)	

[a] Values obtained using elemental combustion data and av mol wt data (vapor-phase osmometry); values in ( ) were obtained directly from ^{1}H and ^{13}C NMR data.
[b] Aromaticity parameter (f_a), the ratio of the aromatic carbons to the total number of carbons.

rings present for this sample. The $(H/C)_{al}$ ratios for these samples range from 1.44–2.02 which suggest the importance of condensed aliphatic ring systems, (e.g., Tetralin, decahydronaphthalene, etc.). Linear and branched alkyl groups are few in these samples based on the $(H/C)_{al}$ ratio and inspection of the ^{1}H spectral region of interest (0.7–1.8 ppm).

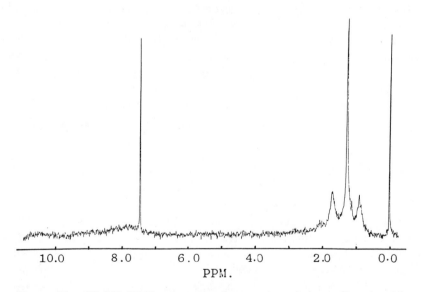

Figure 10. ^{1}H FT NMR spectrum of a portion of Amax Fraction 10 (tubes #38–42, see Figure 3) in d-chloroform with hexamethyldisiloxane reference

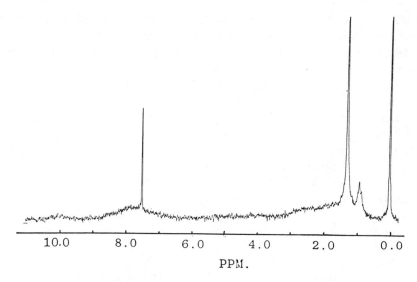

Figure 11. *¹H FT NMR spectrum of a portion of Amax Fraction 20
(tubes #54–58, see Figure 3) in d-chloroform with hexamethyldisiloxane
reference*

The data reported in Table V are generally consistent with the
values presented in our preliminary report (*8*) and earlier work (*6*) with
one exception. The $(H/C)_{al}$ ratio and aromaticity parameter (f_a) for the
Amax 30 sample are higher than reported previously. The reason for this

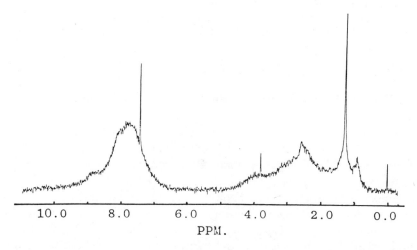

Figure 12. *¹H FT NMR spectrum of a portion of Amax Fraction 30
(tubes #60–64, see Figure 3) in d-chloroform with hexamethyldisiloxane
reference*

discrepancy is twofold: (1) the sample homogeneity problem discussed previously for this sample directly influences the direct ^{1}H and ^{13}C NMR determination of the $(H/C)_{al}$ parameter (e.g., the $(H/C)_{al}$ ratio varies from $\simeq 1.4$–2.0 for the high and low (g_r/g_s) regions discussed in Figure 9, respectively); and (2) the Amax 30 sample used in our preliminary report was contaminated partially with the higher molecular weight fractions (20 and 10). Furthermore, the relatively low occurrence of aliphatic carbon for this sample $(f_a = 0.77)$ and the Monterey 30 sample $(f_a = 0.77)$ corresponds to the lower limits of detection for our ^{13}C NMR instrument. The importance of the $(H/C)_{al}$ parameter is crucial in ascertaining whether highly condensed aliphatic ring systems (e.g., adamantyl) occur in these SRC samples. Unfortunately, solubility limitations prevented ^{13}C NMR examination of Fractions 20 and 10. It is undoubtedly true that the solubility and sample homogeneity problems represent the most formidable limitations in ^{13}C NMR studies of this type.

Perhaps pertinent to the question of aliphatic ring systems ^{1}H FT NMR examination of the first 5-mL cuts in Fractions 10, 20, and 30 for the Amax sample indicate that the aromatic hydrogen content decreases in progressing from Fraction 30 to 10. The ^{1}H spectra for these fractions are presented in Figures 10, 11, and 12 for the d-chloroform-soluble portion of these fractions. The extremely low aromatic hydrogen to total hydrogen ratio for Fraction 10 ($\simeq 0.1$) relative to Fraction 30 (~ 0.5) and the absence of ^{1}H signals in the region above 2 ppm where aliphatic hydrogen α to aromatic rings resonate normally is indeed curious. This trend could be interpreted in terms of either a progression to larger condensed aromatic rings and/or a greater preponderance of aliphatic skeletal carbon for the higher molecular weight fractions (10 and 20). For either possibility, ^{13}C NMR could potentially answer this question.

Conclusion

The results of the present study suggest that solvent-refined coal samples can be separated preparatively utilizing GPC. In addition, the quantitative ^{1}H and ^{13}C NMR approach provides a reasonable method for chemical characterization of the constituents in SRC fractions in terms of average molecular parameters. However, further work is needed toward refining the NMR approach to allow examination of the higher molecular weight fraction. Sample solubility and homogeneity are undoubtedly major factors limiting the precision and accuracy of the high-resolution NMR approach. Although ^{1}H and ^{13}C NMR provide subdivision of the total carbon and hydrogen present in SRC samples in terms of aromatic and aliphatic groups, further characterization of hetero atom functionality would be equally beneficial.

Acknowledgment

The generous financial support of the Commonwealth of Virginia and the Energy Research and Development Administration (ERDA Contract Number EF 77–X–01–2813) is appreciated. We especially thank D. Welsh and J. Hellgeth for performing many of the chromatography experiments.

Literature Cited

1. Schmid, B. K., *Chem. Eng. Prog.* (1975) **71**, 75.
2. Schiller, J. E., *Hydrocarbon Process.* (1977) 147.
3. Bendoraitis, J. G., Cabal, A. V., Callon, R. B., Stein, T. R., Voltz, S. E., Phase I Report, EPRI, January 1976, Contract EPRI 361–1.
4. Coleman, W. M., Wooton, D. L., Dorn, H. C., Taylor, L. T., *Anal. Chem.* (1977) **49**, 533.
5. Wooton, D. L., Dorn, H. C., *Anal. Chem.* (1976) **48**, 2146.
6. Wooton, D. L., Coleman, W. M., Taylor, L. T., Dorn, H. C., *Fuel* (1977) **58**, 17.
7. Glass, T. E., Dorn, H. C., unpublished results.
8. Coleman, W. M., Wooton, D. L., Taylor, L. T., Dorn, H. C., *Am. Chem. Soc., Div. Fuel Chem., Prepr.* (1977) **22**(5), 78.

RECEIVED October 17, 1977.

4

Analysis of Solvent-Refined Coal, Recycle Solvents, and Coal Liquefaction Products

JOSEPH E. SCHILLER

Grand Forks Energy Research Center, Grand Forks, ND 58201

Several solvent-refined coals, their respective process solvents, and coal liquefaction products have been analyzed by vacuum distillation, adsorption chromatography, and mass spectrometry. Quantitative distillations are accomplished using a Kegelrohr oven with 0.2–0.3-g samples. Column chromatography was performed using alumina as the adsorbent with elution by hexane, toluene, chloroform, and 9:1 tetrahydrofuran (THF)/ethanol. Separated fractions or whole samples were analyzed by conventional GC–MS techniques for identification of major compounds. Carbon number analysis was accomplished using high-resolution mass spectrometry for mass identification and low-voltage mass spectrometry for quantitative analysis.

Research in coal conversion is receiving high national priority. At the Grand Forks Energy Research Center (GFERC), a vigorous program is in progress to study product composition, measure reaction kinetics, and evaluate reactor design for the CO–steam process for low-rank coal liquefaction. A batch autoclave capable of hot charging and timed sampling of slurry and gas is used for reactor research. Changes in slurry composition are seen by examination of the small (1-g) samples taken during the run. In this study, analytical methods used for CO–steam product characterization at GFERC were applied to solvent-refined coals (SRC) and recycle solvents, as well as to coal liquefaction products. The utility of these analytical methods was to be demonstrated, and the results were to provide information on the effects of process variables.

Solvent refining (*1*) is a procedure to remove sulfur and ash from coal and increase the heating value. Pulverized coal is dispersed in a

solvent and allowed to react with hydrogen at high pressure and temperature. The solubilized coal solution is filtered to remove ash and unconverted coal, and the solvent is separated by vacuum distillation. The
vacuum bottoms from this process is the SRC, and solvent is recycled.
Several pilot plant scale operations using this process (*2, 3, 4*) are in
existence at present in the United States.

Some coal liquefaction methods, such as the Synthoil (*5*) process
and the CO–steam (*6*) process, are similar to solvent refining in their
approach, but more severe conditions or a catalyst are used to give a
fluid product. In the Synthoil reaction, bituminous coal is pulverized,
dispersed in a vehicle oil, and hydrogenated in a packed tubular reactor
with or without added catalyst. The CO–steam process uses lignite coal
and less expensive synthesis gas. The intended product is a heavy liquid
fuel having ash, sulfur, and nitrogen contents sufficiently low to avoid
stack-gas cleaning. Reactor temperature and pressure are normally 400°–
450°C and 4000 psi, respectively.

Analysis of coal-derived distillates is done often by modified petroleum analytical procedures. In other studies (*7, 8, 9*) as in this one, mass
spectrometry is the primary tool for characterizing the complex mixtures
encountered. Recycle solvents and distillates were subjected to high-resolution mass spectrometry (MS) for compound identification and low-
voltage MS for quantitative measurement of components (*10*).

Process solvents, SRC's, and liquefaction products also were examined
by column chromatography (*11*). The sample was dissolved in chloroform or THF, pre-adsorbed on neutral alumina (*12*), and eluted from
neutral alumina to give saturates, aromatics, and three polar fractions.
The saturate fraction was analyzed quantitatively by mass spectrometry,
and compound identification in distillates and solvents was confirmed by
combined GC–MS or high-resolution MS analysis of column chromatography fractions.

Microdistillation with mass spectral analysis of the distillate yielded
valuable information about the SRC's studied. Although only 2–17%
of the SRC's were volatile under the conditions used, the nature of the
distillate defined the completeness of process solvent separation, solvent
separation parameters, and degree of depolymerization of the coal. Also,
the distillate contains stable reaction intermediates between liquid
products and coal itself.

Experimental

Analysis Scheme. Table I lists the materials studied in this work and
process conditions. The methods used to analyze them are as follows:

Table I. Summary Data on Materials Analyzed

Source of Sample	Project Lignite, University of North Dakota, Grand Forks, ND 58202	Pittsburgh & Midway Coal Co., Research & Dev. Dept., Solvent-Refined Coal Pilot Plant, Dupont, WA 98327 (near Tacoma, WA)
Samples Identification	Run M5C SRL[a]	SRC #308
Dissolver/reactor temperature, °F	800	840
Pressure, psig	2500	1486
Reaction time, min	45	60
Feed rate, lb/hr of coal	50	3000
Temperature of solvent flash °F	600[b]	750
Temperature of solvent flash drum, °F	580	550
Pressure of solvent flash, torr	20	50
Type of coal used	ND lignite	KY 14 bit.
Additional materials analyzed	FS 120 and run M5 inventory[c] solvent	#307 recycle solvent[d]

[a] Ash and unconverted coal were not removed from the product prior to analysis.
[b] Product slurry was not filtered prior to vacuum flash distillation.
[c] Totally coal-derived solvent was not achieved during the run, but solvent inventory at the end of the run is estimated to be 60% coal-derived.
[d] #307 recycle solvent was made during preparation of SRC #308.

1. CO–steam and Synthoil.
   A. Microdistillation.
   B. Mass spectrometry of distillate.
   C. Column chromatography.
2. SRC and SRL (solvent-refined lignite).
   A. Microdistillation.
   B. Mass spectrometry of distillates, where at least 10% was distillable.
   C. Column chromatography.
3. Refining solvents.
   A. Mass spectrometry of saturates and of the unseparated solvent.
   B. Column chromatography to determine saturates.

Analytical Methods

Microdistillation. Microdistillation of samples was performed using a Buchi/Brinkman Kugelrohr apparatus operated at 250°C and 1 torr.

in Grand Forks Energy Research Center Study

Pittsburgh & Midway Coal Co., Research & Dev. Dept., 9009 W. 67th St., Merriam, KS 66202		*Southern Services, Inc., P.O. Box 2625, Birmingham, AL 35202 (pilot plant at Wilsonville, AL)*			*ERDA: Pittsburgh Energy Research Center, 4800 Forbes Ave., Pittsburgh, PA 15213*	
SRC #122	*SRC #115R'*	*Amax SRC*	*Ill. #6 SRC*	*Pitt. #8 SRC*	*CO–Steam Product*	*Synthoil Product*
850	850	826	823	859	795	840
1340	2000	2480	1740	1750	4000	4000
18	120	43	45	57	60	30/2[g]
100	100	373	298	476	4	25
—	—	560	604	599	—	—
480[e]	480[e]	—	—	—	—	—
3	3	70	70	70	—	—
KY 14 bit.	KY 14 bit.	WY subbit.	IL bit.	Pitt. seam bit.	ND lignite	WV bit.
#122 recycle solvent	#115R recycle solvent	AMax recycle solvent	Burning Star recycle solvent	Pitt. #8 recycle solvent	—	—

[e] Solvent was removed by batch distillation.
[f] SRC 115R was made using vigorous conditions and product recycle to generate excess solvent.
[g] The slurry was in a preheater for 30 min and in a tubular reactor packed with silica beads for 2 min.

Sample size was 0.2–0.3 g. Distillate was collected at room temperature, and the amount of distillate and residue were determined by weighing the collection vessel and the distillation pot before and after distillation.

Mass Spectrometry (MS). The instrument was an AEI MS 30 (single-beam) high-resolution mass spectrometer interfaced with a DS–50 data system. The technique used to obtain quantitative data from mass spectra was as follows. Approximately 1 mg of distillate, solvent, or coal liquefaction product was introduced into the mass spectrometer using a heated glass inlet system operating at 325°C. Source temperature was 300°C, the pressure in the ion source was 1.5–2.0 $\times$ 10^{-6} torr, and the ionizing voltage was 70. A high resolution spectrum (8000–9000 resolving power) was obtained so that peaks could be identified by their molecular formulas. Then the ionizing voltage was lowered to 10 V effective (*13*) and the resolution decreased to 1500, and at least six low-voltage spectra were taken. The intensities from these spectra were averaged, and calibration with hydrocarbons, oxygen compounds, and nitrogen compounds was used to calculate concentrations from the low-voltage intensity data. Only a single introduction of sample was required for both high resolution and low-voltage mass spectral analysis.

Column Chromatography. Column chromatography of the solvent refining or liquefaction samples was done in a manner described previously (*11*). The sample (.2 g) was dissolved in THF or chloroform and pre-adsorbed on 2 g of neutral alumina. The solvent was removed under vacuum, and the alumina with sample was added to an 11-mm o.d. glass column containing an additional 6 g of neutral alumina (Activity I). Separation into saturates, aromatics, and various polar materials was done by elution with hexane, toluene, chloroform (two fractions), and 9:1 THF/ethanol.

Gas Chromatography–Mass Spectrometry (GC–MS). Materials were analyzed by GC–MS using a Varian 2740 gas chromatograph coupled with a DuPont 21–491B mass spectrometer. The GC column was 6 ft x 2-mm i.d. glass packed with 3-% SE–30 on Varaport 30. The flow rate of helium carrier gas was 30 mL/min, and the temperature was programmed from 70–270°C at 6°C/min. GC peak areas were determined using a Spectra Physics System I Computing Integrator and response factors were measured with appropriate pure standards. Mass spectra for identification of GC peaks were obtained using computer control of MS scanning with the AEI DS–50 data system. The ion source was 250°C, and the ionizing voltage was 70 V.

Results and Discussion

Column Chromatography. A column chromatographic method was desired which would separate CO–steam samples into fractions which could be further analyzed by mass spectrometry. We desired a method which would be rapid and simple and would not require a highly skilled laboratory technician to perform it. We wanted to obtain only a few fractions, because many samples were to be analyzed and a multiplicity of effort would develop for every fraction obtained by column chromatography. Many of the procedures used for separation of petroleum were not suitable, either because they were too time consuming or because they required a sample soluble in hexane, or both. The column chromatography procedure used in our laboratory isolated one fraction of saturated hydrocarbons and a fraction containing aromatic hydrocarbons. Both of these fractions were subjected routinely to mass spectrometric analyses and GC–MS analysis for both compound identification and measurement of concentration of individual compounds. MS was used to analyze the saturates for compound types present, such as paraffins, cycloparaffins, dicycloparaffins, etc. and aromatics for naphthalenes, phenanthrenes, etc.

In contrast to petroleum distillates and residuals, the coal-derived solids and liquids studied in this work contained only very small concentrations of saturated hydrocarbons. The SRC's had 1–2% saturated hydrocarbons, the CO–steam and Synthoil products had ~ 5%, and coal-derived SRC recycle solvents had ~ 10% saturates. The petroleum-derived FS–120 used by Project Lignite as start-up solvent for SRL operation contained ~ 20% saturates. The size of the aromatic fractions showed similar trends. The SRC's studied generally contained 10% or less of hydrocarbons with less than six condensed rings. The coal lique-faction products had about 25% aromatics, and SRC recycle solvents contained typically 60% aromatics. From these data, it is clear that the final products of coal liquefaction (SRC, SRL, CO–steam, and Synthoil) are mainly hetero atomic and/or polymeric in nature.

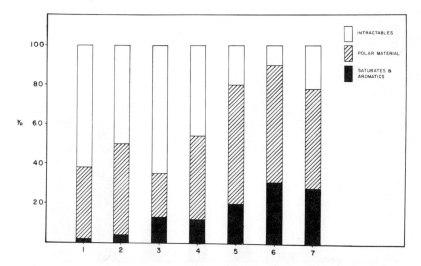

Figure 1. Column chromatographic data for SRC's and coal liquefaction products: 1. P & M 308; 3. P & M 122; 3. Project Lignite SRL–M5C; 4. Wilsonville SRC–Pitt #8; 5. P & M 115R; 6. CO–Steam Product; 7. Synthoil

In addition to the pure saturates and aromatics that were isolated for further analysis, the column chromatography data itself was mean-ingful. As seen in Figure 1, a different elution profile is obtained depending on how completely the coal has been depolymerized or how completely the solid has been converted to liquid-type products. A material that is eluted completely by the column chromatography pro-cedure has reacted further than another material where only 50, 70, or 80% of it is eluted. Therefore, we could get a measure of the degree of conversion by the total amount of elutable material in the sample and also the size of the various fractions produced. The accuracy and precision of the size of column chromatography fractions range from about 2% absolute for small fractions up to 5% absolute for large fractions.

Microdistillation. A microdistillation procedure was developed which would allow determination of volatile components in 0.2–0.3-g samples. Again, our object in developing these methods was to look at timed samples from a batch autoclave, and these samples were ~ 1 g in size. Values for percent distillable and percent residue are accurate and reproducible within ~ 3%. In addition to using this procedure routinely on CO–steam liquefaction samples, it was applied to solvent-refined coal, and some interesting differences appeared (Figure 2). Two of

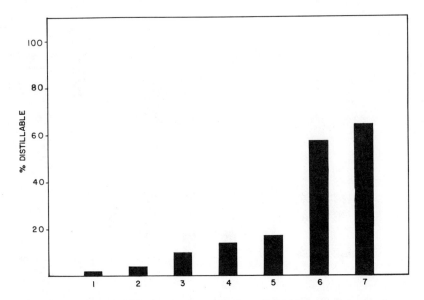

Figure 2.   Microdistillation data for SRC's and coal liquefaction products: 1. P & M 308; 2. P & M 122; 3. Project Lignite SRL–M5C; 4. Wilsonville SRC–Pitt #8; 5. P & M 115R; 6. CO–Steam Product; 7. Synthoil

the solvent-refined coals studied in this work showed very low concentrations of distillate. The solvent-refined coal from the Tacoma pilot plant and from Gulf Research at Merriam, Kansas showed 2% and 4%, respectively, of distillable material under the conditions used. Other solvent-refined coals and lignites had up to 17% distillable. This broad range reflects variations in the process condition used for separation of recycle solvent from the vacuum bottoms (SRC). These differences are discussed in more detail in the section on comparison of solvent-refined coals. The technique of microdistillation was also useful in isolating volatiles from solvent-refined coal from subsequent analysis by MS or GC–MS. If the volatiles were not isolated, the analyses by GC or MS were more difficult because a larger sample had to be used and the inlets of the GC or MS would become contaminated with the nonvolatile material from the SRC samples.

Mass Spectrometry (MS). The MS methods described here resemble the published procedures used in petroleum and coal-liquids analysis, with the exception that the MS 30 used in this work was operated both in low resolution and high resolution on the same sample. The slits were changed for analysis of samples by both techniques before it was pumped away for succeeding analysis of the next sample. This was accomplished in the following way. Samples are introduced into the mass spectrometer in the conventional fashion. Then at high resolution (8000–9000 resolving power) and high ionizing voltage (70 V), perfluorkerosene and the sample were introduced simultaneously into the ion source to give a spectrum that would allow identification of major peaks from their exact mass. The resolution and the ionizing voltage were decreased then to 1500 and 10.0 V, respectively. This had the effect of increasing sensitivity by increasing the intensity of the ion beam at low resolution while the voltage was decreased so that only parent ions were produced from the sample. After several scans at low voltage, the intensities for the masses seen during each scan were averaged. This gave accurate quantitative data for the sample. High-resolution MS was a very rapid way of getting qualitative information about the sample, and the combination of these two techniques is a method for characterization of previously unencountered sample types.

We found this technique also valuable for looking at recycle solvents and volatiles from SRC's. For major constituents, GC–MS was used to confirm high-resolution MS identification and to determine specifically which compounds of possible isomers were contributing at various molecular formulas. Liquefaction processes using no catalyst (such as CO–steam) or solvent-refining processes give liquids which are relatively simple mixtures. For the samples encountered, ~ 15–20 compounds in these mixtures accounted for 50–75% of the total. This is opposite the trend observed in petroleum where a great many compounds are seen and where the samples contain perhaps thousands of individual compounds. Indeed coal liquids are complex mixtures, but it does not appear that liquids produced by noncatalytic or mild hydrogenative procedures are as broad in composition as is petroleum. Coal liquids made by the Synthoil process or by coal liquefaction using a high concentration of hydrogen-donor solvent, on the other hand, do contain many more individual compounds than those liquids described here (*14*).

The accuracy of the low-voltage MS method used in our laboratory has been tested by comparison of results with GC analysis of the same samples. Data are shown in Table II for the composition of two coal-

Table II. Comparison of Gas Chromatographic and Mass Spectral Analysis of Two Coal-Derived Oils

Compound	Crowley Coal Tar Oil		Allied Creosote Oil	
	MS (66 masses) %	GC (28 peaks) %	MS (9 masses) %	GC (33 peaks) %
Naphthalene	5.6	8.5	5.3	6.6
Methylnaphthalenes	15.5	20.3	14.1	16.4
Fluorene	6.2	7.0	6.0	6.7
Phenanthrene	16.1	18.8	14.3	17.2
Pyrene and fluoranthene	11.0	12.7	9.2	12.8

derived distillate oils as determined both by low-voltage MS and GC. Values by MS are consistently ∼ 15% (relative) lower than for the same compounds by GC. The GC resolved these mixtures into only 28 and 33 peaks, respectively, while the MS saw 66 and 99 masses. The resolution by GC therefore was not complete, and GC peaks were inflated by certain amounts of unseparated material under all of the observed peaks in the chromatograms. The MS data is lower because of better

Table III. Data from Repetitive Analysis for Percent of Total Ion Intensity of Selected Masses in a Coal Oil Distillate

Mass	Run 1[a]	Run 2	Run 3	Run 4	Run 5	Mean	Standard Deviation	Relative Standard Deviation
94	1.84	1.62	1.77	1.56	1.81	1.72	0.123	7.1%
108	2.42	2.24	2.45	2.32	2.39	2.36	0.084	3.6%
128	27.49	27.07	27.19	28.67	27.04	27.49	0.682	2.5%
132	5.07	4.70	4.97	4.93	4.70	4.87	0.167	3.4%
142	6.03	5.90	5.91	6.18	5.85	5.97	0.133	2.2%
154	8.04	7.82	7.85	8.28	8.05	8.01	0.185	2.3%
156	2.64	2.41	2.46	2.55	2.40	2.49	0.102	4.1%
166	5.15	5.10	5.12	5.19	5.10	5.13	0.038	0.7%
168	10.07	10.03	9.78	10.28	9.82	10.00	0.203	2.0%
178	8.28	8.30	8.06	8.26	8.22	8.22	0.096	1.2%
180	1.66	1.87	1.80	1.79	1.87	1.80	0.086	4.8%
182	4.26	4.33	4.25	4.19	4.26	4.26	0.050	1.2%
192	1.05	0.98	0.98	0.93	1.00	0.99	0.043	4.3%
202	3.65	3.80	3.58	3.50	3.78	3.66	0.129	3.5%

[a] All numbers in the table for Runs 1–5 are percentages of the total ion intensity for each run.

resolution of components, so values from the MS reflect a more accurate analysis. The inaccuracy of GC analysis will increase with more complex samples.

Reproducibility and precision of the low-voltage MS analysis were determined also. Table III shows values from five analyses of the same sample, together with statistical data from the results. Relative standard deviations are generally $< 4\%$, and they increase slightly for components present in very low concentrations.

Solvent-Refined Coal Comparison. The Pittsburgh and Midway process used at the Tacoma pilot plant appears to produce SRC with minimum degradation of the original coal and complete removal of process solvent. There is only 2% distillable at 250°C and 1 torr, showing that the solvent residue left in the vacuum bottoms is very low. A further important point is that the amount of aromatics that appear by column chromatography is extremely low. This low percentage of aromatics indicates that there is little breakdown of the coal structure to aromatic systems with less than five or six condensed rings. That means that the removal of sulfur and ash from the coal during this solvent-refining process is done effectively without chemically depolymerizing the coal to a very great extent. This SRC (#308) contained 0.29% ash and 0.50% sulfur.

Approximately 10% of the Wilsonville SRCs are distillable at 250°C and 1 torr. There is, however, essentially no solvent residue left in the coal, since MS analysis of distillate from the SRC is shown to contain higher molecular weight molecules than make up the recycle solvent. The distillate material in the solvent-refined coal product is composed mainly of phenanthrene/anthracene and pyrene/fluoranthrene, together with lesser amounts of alkylated homologs. Since these three- and four-ring condensed aromatics are not solvent residue, some coals are being depolymerized to produce these compounds.

The Pittsburgh and Midway product from Merriam, Kansas, #122, is quite similar to the Tacoma SRC product, #308. The amount of solvent residue is very low. The solvent in both cases is of moderate molecular weight, and there appears to be very little breakdown of large molecules in the coal to produce either volatile compounds that would be left as residue in the product or to produce aromatic compounds that could be eluted in the column chromatographic analysis.

The Project Lignite SRL contains a moderate percentage of distillable material, and this distillate resembles the process solvent. The mass spectrum of this distillate is essentially a fingerprint of the solvent at mass 178 and above. These data would appear to indicate that some material in the boiling range of the solvent was not separated completely from the SRL. Appearance of distillable compounds in the SRL also might reflect a true difference in product composition, i.e., solvent refining of lignite inherently may give a higher yield of lower molecular weight products than does solvent refining of bituminous coals.

The Pittsburgh and Midway product 115R is somewhat intermediate in properties between a true SRC and a liquefaction product. The amount of distillable material in this SRC is somewhat higher (17%) than the other SRC's. Further, the high concentration of aromatics in this SRC (19%) approaches the values of 23 and 25% aromatics for Synthoil and CO–steam products, respectively. The removal of solvent from the Pittsburgh and Midway 115R product has been effective. The distillable material in this SRC is much higher molecular weight than the recycle solvent. Therefore one again sees depolymerization of very large molecules in the coal to produce heavy aromatics. These compounds that are distillable are heavier than those found in the solvent but still of sufficiently low molecular weight to be volatile under the analysis conditions used.

The Synthoil product and the CO–steam product approach true liquids. Synthoil has a relatively high percentage of distillate (64%), and the CO–steam product is 57% distillable. The amount of saturates and aromatics in these materials is $\sim 30\%$, and the total amount which is eluted by column chromatography is 80–90%.

Acknowledgment

The author wishes to thank the various organizations listed in Table I who supplied samples studied in this work. The author also wishes to thank D. J. Miller and D. H. Neal of the Grand Forks Energy Research Center who performed much of the experimental work.

Reference to specific brands and models is for identification purposes only, and does not represent endorsement by ERDA.

Literature Cited

1. Frank, M. E., Schmid, B. K., "Clean Fuels from Coal," Inst. Gas Technol., Chicago, Symposium papers (1973) 577–600.
2. Interim Report for Research Project #1234, Electric Power Research Institute, Palo Alto, CA, May 1975.
3. Severson, D. E., Souby, A. M., Kube, W. R., U.S. Bur. Mines Inf. Circ. (1974) 8650, 236.
4. Klass, D. L., Chemtech (1975) 5(8), 499.
5. Akhtar, S., Mazzocco, N. J., Weintraub, M., Yavorsky, P. M., Energy Commun. (1975) 1(1), 21.
6. Appell, H. R., Wender, I., Am. Chem. Soc., Div. Fuel Chem., Prepr. (1968) 12(3), 220.
7. Aczel, T., Foster, J. Q., Karchmer, J. H., Am. Chem. Soc., Div. Fuels Chem., Prepr. (1969) 13(1), 8.
8. Peters, A. W., Bendoraitis, J. G., Anal. Chem. (1976) 48, 968.
9. Swansiger, J. T., Dickson, F. E., Best, H. T., Anal. Chem. (1974) 46, 730.

10. Schiller, J. E., *Hydrocarbon Process.* (1977) **1,** 147–152.
11. Schiller, J. E., Mathiason, D. N., *Anal. Chem.* (1977) **49,** 1225.
12. Middleton, W. R., *Anal. Chem.* (1967) **39,** 1839.
13. Johnson, B. H., Aczel, T., *Anal. Chem.* (1967) **39,** 682.
14. Schiller, J. E., Knudson, C. L., in press.

RECEIVED August 5, 1977.

5

Structural Characterization of Solvent Fractions from Five Major Coal Liquids by Proton Nuclear Magnetic Resonance

I. SCHWAGER, P. A. FARMANIAN, and T. F. YEN

University of Southern California, Chemical Engineering Department, University Park, Los Angeles, CA 90007

Structural characterization studies have been made on solvent-fractionated coal liquefaction products by proton nuclear magnetic resonance methods. Coal liquids from the three direct general processes for converting coals to liquid fuels (catalyzed hydrogenation, staged pyrolysis, and solvent refining) have been separated by solvent fractionation into five fractions. The fractions are propane-soluble (oil), propane-insoluble and pentane-soluble (resin), pentane-insoluble and benzene-soluble (asphaltene), benzene-insoluble and carbon disulfide-soluble (carbene), carbon disulfide-insoluble (carboid). Structural parameters such as aromaticity, f_a, the ratio of substitutable edge atoms to total aromatic atoms, H_{aru}/C_{ar}, and the degree of substitution on aromatic rings, σ, provide, in conjunction with analytical and molecular weight data, a series of average molecular properties for the different classes of coal liquefaction products.

It was shown previously that coal liquids from five major demonstration processes (Synthoil, HRI H–Coal, FMC–COED, PAMCO SRC, and Catalytic Inc. SRC) could be separated reproducibly, in high yields (94–99%) into five fractions by solvent fractionation (*1,2*). We now wish to report structural characterization parameters obtained for these materials by the use of proton nuclear magnetic resonance (¹H NMR).

High-resolution ¹H NMR spectrometry was used first by Brown and Ladner for structural characterization of coal pyrolysis products (*3*). Other workers have extended this type of analysis to coal extracts (*4*) and coal hydrogenation products (*5,6,7*). A recently published com-

0-8412-0395-4/78/33-170-066$05.00/1

parison between f_a values determined from ^{13}C NMR spectra, and those estimated from ^1H NMR data demonstrated that the use of the ^1H NMR method is reasonably reliable for coal-derived materials (8). The structural parameters calculated are: f_a, fraction of total carbon which is aromatic carbon; H_{aru}/C_{ar}, ratio of substitutable edge atoms to total aromatic atoms, σ, fraction of the available aromatic edge atoms occupied by substituents; R_S, number of substituted aromatic ring carbons; n, number of carbon atoms per saturated substituent; C_A, total number of aromatic carbon atoms; R_A, number of aromatic rings.

Experimental

The general solvent fractionation scheme and the composition and analysis of the coal liquid fractions were described previously (1, 2).

^1H NMR spectra were run on a Varian T–60 spectrometer. The solvent used ordinarily was 99.8% $DCCl_3 + 1\%$ tetramethylsilane (TMS), but because of solubility limitations, the carboids were run in 99.5% dimethylformamide (DMF)–d_7 or pyridine–d_5. The concentrations were generally 10–20 wt/vol %.

Trimethylsilyl ethers of asphaltenes were synthesized by refluxing the asphaltene with excess 1,1,1,3,3,3-hexamethyldisilazane (HMDS) in dry tetrahydrofuran (THF). After removal of solvent by rotary evaporation and decanting of the unreacted HMDS, the product was dried under vacuum to constant weight. The number of trimethylsilyl groups introduced by silylation of the asphaltenes was determined by silicon analysis (9).

NMR analysis was carried out using modified Brown–Ladner equations (3, 7, 10). These average structural parameter equations are given in Table I.

Results and Discussion

The solvent fractionation scheme for separating coal liquid products into five fractions: propane-soluble (oil); propane-insoluble and pentane-soluble (resin); pentane-insoluble and benzene-soluble (asphaltene); benzene-insoluble and carbon disulfide-soluble (carbene); and carbon disulfide-insoluble (carboid) was described previously (1, 2).

The ^1H NMR spectrum of a typical coal asphaltene is shown in Figure 1. The centers of absorption for different types of protons are marked with arrows: $\delta = 7.25$, $H_{ar} =$ aromatic protons; $\delta = 2.40$, $H\alpha =$ protons α to aromatic rings; $\delta = 1.58$, $H_N =$ naphthenic protons; $\delta = 1.25$, $H_R =$ methylenic protons; $\delta = 0.9$, $H_{SMe} =$ saturated methyl protons. Brown–Ladner analysis requires that the three areas of absorption centered at $\delta = 7.3$, 2.4, and 1.2 ppm be assigned to aromatic ring hydrogens (H_{ar}), aliphatic hydrogens adjacent to aromatic rings ($H\alpha$), and aliphatic hydrogens not adjacent to aromatic rings (H_o). The separation point

Table I.   Modified Brown–Ladner Equations[a]

$$f_a = \frac{\dfrac{C}{H} - \dfrac{H_\alpha^*}{x} - \dfrac{H_o^*}{y}}{\dfrac{C}{H}} = \text{fraction of total carbon which is aromatic carbon}$$

$$\frac{H_{aru}}{C_{ar}} = \frac{\dfrac{H_\alpha^*}{x} + H_{ar}^* + \dfrac{0-0_{OH}}{H}}{\dfrac{C}{H} - \dfrac{H_\alpha^*}{x} - \dfrac{H_o^*}{y}} = \text{ratio of substitutable aromatic edge atoms to total aromatic atoms}$$

$$\sigma = \frac{\dfrac{H_\alpha^*}{x} + \dfrac{0}{H}}{H_{ar}^* + \dfrac{H_\alpha^*}{x} + \dfrac{0-0_{OH}}{H}} = \text{fraction of the available aromatic edge atoms occupied by substituents}$$

$$R_S = \sigma C_A \frac{H_{aru}}{C_{ar}} = \text{number of substituted aromatic ring carbons}$$

$$n = \frac{H_o^*}{H_\alpha^*} + 1 = \text{number of carbon atoms per saturated substituent}$$

$$C_A = \frac{f_a (C) (M)}{100} = \text{total number of aromatic carbon atoms}$$

$$R_A = C_A \left(\frac{1 - H_{aru}/C_{ar}}{2} \right) + 1 = \text{number of aromatic rings}$$

C = mol % carbon

H = mol % hydrogen

0 = mol % oxygen

0_{OH} = mol % phenolic oxygen

H_{ar}^* = mol fraction aromatic hydrogen

H_α^* = mol fraction hydrogen α to aromatic ring

H_o^* = mol fraction of aliphatic hydrogen not α to aryl ring

x = average ratio of hydrogen to carbon on carbons α to aryl ring

y = average ratio of hydrogen to carbon on aliphatic carbons not α to aryl ring

[a] Assumptions: All oxygen is attached to aryl rings in either phenol or aryl ether groups. All phenolic hydrogens found by the TMS derivative method are under the aryl absorption. $x = y = 2$.

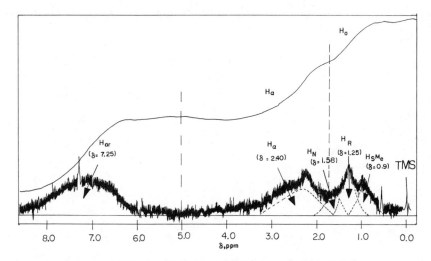

Figure 1. *¹H NMR spectrum of PAMCO SRC asphaltene*

between the H_α and H_o protons was chosen at $\delta = 1.73$ ppm. Because hydrogen-bonded phenolic OH resonances also are believed to be shifted under the aromatic envelope (*11, 12*), it is necessary to determine the percent of phenolic OH in order to correct the H_{ar} value. Therefore the silylation procedure of Friedman et al. (*9*) was carried out with all asphaltenes. The silylated asphaltenes were examined by IR, and found to contain essentially no asphaltene OH bands. The asphaltene N–H band at ≈ 3450 cm^{-1} remained, and trimethylsilyl ether bands appeared at 1250 and 850 cm^{-1}.

The silylation results are presented in Table II. The results indicate that asphaltenes produced in SRC processes (PAMCO and Catalytic Inc.) contain practically all of their oxygen as phenolic oxygen. However, asphaltenes from catalyzed hydrogenation (Synthoil and HRI H–Coal) and staged pyrolysis (FMC–COED) contain only 51–61% of

Table II. Analysis of Silylated Asphaltenes

Asphaltene	% Si in Silylated Asphaltene	% OH in Silylated Asphaltene[a]	% O in Starting Asphaltene[b]	% OH/O$_{tot}$ in Starting Asphaltene
HRI H–Coal	4.11	2.62	4.96	53
Synthoil	2.82, 3.22	1.74, 2.00	3.93	44, 51
FMC–COED	6.31	4.30	7.11	61
PAMCO SRC	6.58	4.52	4.68	97
Cat. Inc. SRC	6.17	4.19	4.58	92

[a] Using method of Friedman, et al. (*9*).
[b] Determined by difference.

Table III. Average Molecular

Molecular Formula	FMC–COED $C_{19.11}H_{26.03}N_{0.08}O_{0.48}S_{0.09}$	Synthoil $C_{17.77}H_{23.28}N_{0.1}O_{0.25}S_{0.03}$
Mol wt[b]	267	243
H_{ar}*	0.16	0.17
H_{α}*	0.30	0.35
H_o*	0.54	0.48
f_a	0.43	0.46
H_{aru}/C_{aru}	1.03	1.01
σ	0.51	0.52
R_S	4.4	4.3
n	2.8	2.4
C_A	8.3	8.2
R_A	0.9	1.0

[a] Brown–Ladner Method $x = y = 2$.

the oxygen present as phenolic oxygen. Since a different coal was used in each of the SRC processes (West Kentucky, hvBb, and Wyoming sub-bituminous) and similar coals were used in the Synthoil, FMC–COED, and PAMCO processes (West Kentucky, hvAb, and hvBb), these results do not appear to be a consequence of the coal used in the process. The results could be explained by assuming that the relatively mild SRC processes permit retention of phenolic functions in the asphaltenes, whereas the more vigorous Synthoil, H–Coal, and FMC–COED processes result in partial conversion of phenolic groups to ether groups in the asphaltenes produced in these processes (13, 14).

The average molar properties of the soils, resins, asphaltenes, and carboids are presented in Tables III–VI. The oils, with the exception of the SRC products, are fairly aromatic species having from 43–54% of

Table IV. Average Molecular

Molecular Formula	FMC–COED $C_{22.55}H_{23.56}N_{0.25}O_{0.22}S_{0.73}$	Synthoil $C_{22.18}H_{23.70}N_{0.28}O_{0.67}S_{0.01}$
Mol wt[b]	325	305
H_{ar}*	0.28	0.29
H_{α}*	0.37	0.35
H_o*	0.35	0.36
f_a	0.63	0.62
H_{aru}/C_{aru}	0.87	0.84
σ	0.47	0.41
R_S	5.8	4.8
n	2.0	2.0
C_A	14.1	13.8
R_A	1.9	2.1

[a] Brown–Ladner Method $x = y = 2$.

Properties of Oils[a]

HRI H–Coal	Catalytic Inc. SRC	PAMCO SRC
$C_{16.58}H_{20.25}N_{0.07}O_{0.09}S_{0.00}$	$C_{20.02}H_{18.32}N_{0.08}O_{0.18}S_{0.05}$	$C_{23.00}H_{22.40}N_{0.09}O_{0.41}S_{0.01}$
222	264	307
0.24	0.48	0.40
0.33	0.28	0.37
0.42	0.24	0.23
0.54	0.76	0.71
0.94	0.75	0.82
0.41	0.24	0.34
3.5	2.7	4.6
2.3	1.9	1.6
8.9	15.3	16.3
1.3	2.9	2.5

[b] VPO in benzene.

the carbon as aromatic carbon. They have approximately one aromatic ring per average molecule, and contain only small amounts of the hetero-atoms nitrogen, oxygen, and sulfur. These molecules are moderately substituted with 41–52% of the available aromatic edge carbons being substituted. The average number of carbon atoms per saturated substituent ranges from 2.3–2.8. The oil fractions from the two SRC fractions contain average molecules which are larger and more aromatic than those found in the other oil fractions from the pyrolysis and catalyzed hydrogenation processes. This suggests less breakdown of coal and coal liquefaction intermediates to small molecules in SRC processes than in the other types of processes studied. Another contributing factor in the higher aromaticity and aromatic ring size in SRC oil and resin fractions is that SRC products are vacuum flash tower bottoms, and might have had a

Properties of Resins[a]

HRI H–Coal	Catalytic Inc. SRC	PAMCO SRC
$C_{20.10}H_{18.94}N_{0.22}O_{0.38}S_{0.01}$	$C_{27.71}H_{24.57}N_{0.22}O_{0.61}S_{0.00}$	$C_{26.07}H_{25.48}N_{0.22}O_{0.56}S_{0.03}$
270	370	352
0.40	0.43	0.37
0.35	0.30	0.39
0.25	0.27	0.25
0.72	0.75	0.69
0.77	0.71	0.82
0.33	0.29	0.38
3.7	4.3	5.5
1.7	1.9	1.6
14.4	20.8	17.9
2.7	4.0	2.6

[b] VPO in benzene.

Table V.   Average Molecular

	FMC–COED	*Synthoil*
Molecular Formula	$C_{25.85}H_{24.26}N_{0.46}O_{1.49}S_{0.30}$	$C_{40.87}H_{35.05}N_{0.69}O_{1.36}S_{0.07}$
Mol wt[b]	375	560
H_{ar}^{*}	0.248	0.312
H_{OH}^{*}	0.042	0.018
H_{α}^{*}	0.45	0.42
H_{o}^{*}	0.26	0.25
f_a	0.67	0.71
H_{aru}/C_{aru}	0.75	0.68
σ	0.54	0.44
R_S	7.0	8.7
n	1.6	1.6
C_A	17.3	29.1
R_A	3.1	5.7

[a] Brown–Ladner Method $x = y = 2$.

higher percentage of more saturated oil and resin molecules removed in the distillation.

The resins show an increase in heteroatoms and small decreases in substitution, and in the number of carbon atoms per saturated substituent. Again SRC resins appear to have larger percentages of aromatic carbons and a larger average aromatic ring system.

Average coal asphaltene molecules are highly aromatic species having 67–80% of the carbon as aromatic carbon. These molecules are moderately substituted with 32–54% of the available aromatic edge carbons being substituted. The average number of carbon atoms per saturated substituent is 1.4–1.6. The number of aromatic rings ranges from 3.1 to 6.2.

Table VI.   Average Molecular

	FMC–COED	*Synthoil*
Molecular Formula	$C_{25.76}H_{22.26}N_{0.55}O_{2.98}S_{0.23}$	$C_{69.04}H_{53.37}N_{1.10}O_{1.34}S_{0.61}$
Mol wt[b]	394	938
H_{ar}^{*}	0.43	0.49
H_{α}^{*}	0.39	0.35
H_{o}^{*}	0.18	0.16
f_a	0.76	0.80
H_{aru}/C_{aru}	0.86	0.66
σ	0.43	0.29
R_S	7.3	10.6
n	1.5	1.5
C_A	19.5	55.5
R_A	2.3	10.5

[a] Brown–Ladner Method $x = y = 2$.

Properties of Asphaltenes[a]

$HRI\ H-Coal$	$Catalytic\ Inc.\ SRC$	$PAMCO\ SRC$
$C_{35.59}H_{28.86}N_{0.50}O_{1.52}S_{0.10}$	$C_{36.60}H_{27.57}N_{0.44}O_{1.31}S_{0.02}$	$C_{40.81}H_{33.70}N_{0.63}O_{1.65}S_{0.11}$
492	483	563
0.432	0.461	0.373
0.026	0.039	0.047
0.36	0.34	0.38
0.18	0.16	0.20
0.78	0.80	0.76
0.69	0.67	0.67
0.35	0.32	0.39
6.7	6.2	8.1
1.5	1.5	1.5
27.8	29.4	31.1
5.3	5.7	6.2

[b] Average of infinite dilution values in THF and benzene.

Average molecules of carboids (Table VI) generally have the largest molecular weights and the largest aromatic ring systems. They contain the most heteroatoms found in any of the solvent fractions. The degree of substitution is less than in the asphaltenes, and the number of carbon atoms per saturated substituent is about the same as is found in the asphaltene fractions.

The structural parameter H_{aru}/C_{ar}, which gives the ratio of substitutable edge atoms to total aromatic atoms, is important because it is a measure of the average aromatic ring system which is independent of molecular weight measurements. Figure 2 shows the H_{aru}/C_{ar} values for some typical 1–7 cata-condensed aromatic ring systems. When the $H_{aru}/$

Properties of Carboids[a]

$HRI\ H-Coal$	$Catalytic\ Inc.\ SRC$	$PAMCO\ SRC$
$C_{73.52}H_{49.55}N_{1.11}O_{4.21}S_{0.31}$	$C_{74.45}H_{48.22}N_{0.70}O_{4.44}S_{0.11}$	$C_{42.66}H_{29.29}N_{0.43}O_{2.81}S_{0.22}$
1020	1026	600
0.52	0.51	0.485
0.31	0.34	0.31
0.17	0.15	0.205
0.84	0.84	0.82
0.61	0.59	0.61
0.33	0.34	0.34
11.9	12.6	7.4
1.5	1.4	1.7
61.8	62.7	35.1
12.9	14.0	7.8

[b] VPO in DMF.

C_{ar} values of the coal liquid fractions are plotted vs. the atomic C/H ratios of the same fractions (Figure 3), a smooth decrease in H_{aru}/C_{ar} vs. C/H is obtained. This indicates that the increase in C/H values is caused by an increase in the size of the average polynuclear condensed aromatic ring systems in going from oils to carboids. The number of cata-condensed aromatic rings representative of the H_{aru}/C_{ar} values is shown on the left side of the figure. This comparison indicates that oils contain ~ 1 aromatic ring; resins, ~ 2–3 aromatic rings; asphaltenes, ~ 3–5 aromatic rings; and carboids ~ 4–7 aromatic rings per aromatic ring system.

It is interesting to note that the atomic C/H values of coals fall generally into the range of 1.16–1.26 which would correspond to ~ 3 aromatic rings per average aromatic unit in coals if coal followed the

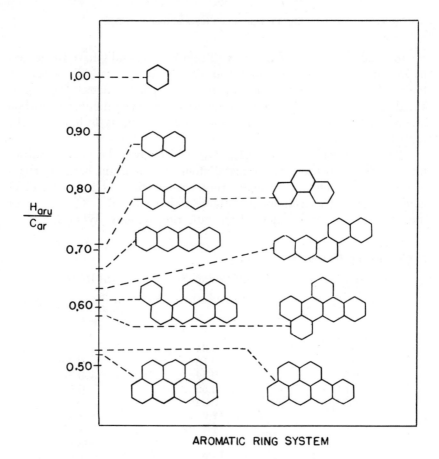

AROMATIC RING SYSTEM

Figure 2.   Comparison of H_{aru}/C_{ar} values for cata-condensed aromatic ring systems

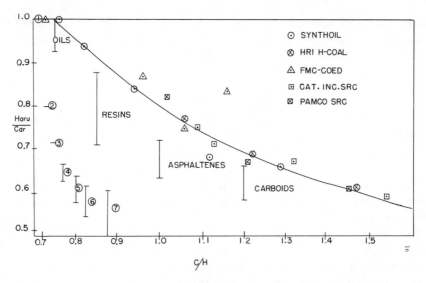

Figure 3.   Variation in average aromatic ring size vs. atomic C/H ratio of coal liquid fractions

curve. This further suggests that carboids, which are more unsaturated than coal, and which apparently have a larger size aromatic unit than coal, are formed either by dehydrogenation of coal or coal liquefaction intermediates, or by polymerization of such reactive coal depolymerization species which have not been stabilized by addition of hydrogen. A general mechanism consistent with such observations for coal liquefaction product is:

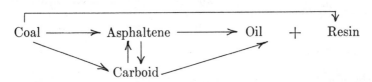

Conclusions

Solvent fractions of coal liquids such as oils, resins, asphaltenes, and carboids may be separated because of their different polarity, molecular weights, and degree of aromatic character. The composition of coal liquids appears to be more closely related to the liquefaction process used rather than to the type of coal used. For example, oils and resins produced in SRC processes appear to have higher molecular weight, more aromatic average molecules than those contained in either pyrolysis or catalyzed hydrogenation coal liquids. This suggests less breakdown of coal and coal liquefaction intermediates to smaller, more saturated molecules in SRC processes than in the other processes studied. Asphaltenes produced in SRC processes appear to contain practically all of their oxygen (92–97%) as phenolic oxygen, whereas asphaltenes from catalyzed hydrogenation and staged pyrolysis contain only 51–61% of the oxygen present as phenolic oxygen. These observations may be explained by assuming that the relatively mild SRC processes permit retention of phenolic functions in the asphaltenes, whereas the more vigorous hydrogenation and staged-pyrolysis processes result in partial conversion of phenolic groups into ether groups.

A relatively smooth decrease in the structural parameter H_{aru}/C_{ar} was observed as a function of the increase in atomic C/H values for the solvent fractions in going from oils to carboids. This suggests that the increase in unsaturation is a consequence of the increase in the size of the average polynuclear-condensed aromatic ring system. Comparison of average H_{aru}/C_{ar} values for coal liquid solvent fractions with those values calculated for known cata-condensed aromatic ring systems suggests that the average aromatic ring system contains ~ 1 aromatic ring for oils, ~ 2–3 aromatic rings for resins, ~ 3–5 aromatic rings for asphaltenes, and ~ 4–7 aromatic rings for carboids.

Acknowledgment

The authors wish to thank the FMC Corporation, Hydrocarbon Research Inc., Catalytic Inc., PAMCO, and the Pittsburgh Energy Research Center of ERDA for generously supplying samples of their coal liquid products. The work described here was sponsored by the United States Energy Research and Development Administration under Contract No. E(49–18) 2031.

Literature Cited

1. Schwager, I., Yen, T. F., *Am. Chem. Soc., Div. Fuel Chem., Prepr.* (1976) **21**(5), 199.
2. Schwager, L., Yen, T. F., *Fuel* (1978) **57**, 100.
3. Brown, J. K., Ladner, W. R., *Fuel* (1960) **39**, 87.
4. Retcofsky, H. L., Friedel, R. A., *in* "Spectrometry of Fuels," R. A. Friedel, Ed., Chap. 6–8, Plenum, New York, 1970.
5. Maekawa, Y., Ueda, S., Hasegawa, Y., Nakata, Y., Yakoyama, S., Yoshida, Y., *Am. Chem. Soc., Div. Fuel Chem., Prepr.* (1975) **20**(3), 1.
6. Yakoyama, S., Bodily, D. M., Wiser, W. H., *Am. Chem. Soc., Div. Fuel Chem., Prepr.* (1976) **21**(7), 77, 84.
7. Woolsey, N., Baltisberger, R., Klabunde, K., Stenberg, V., Kaba, R., *Am. Chem. Soc., Div. Fuel Chem., Prepr.* (1976) **21**(7), 33.
8. Retcofsky, H. L., Schweighardt, F. K., Hough, M., *Anal. Chem.* (1977) **49**, 585.
9. Friedman, S., Zahn, L., Kaufman, M., Wender, I., *Fuel* (1961) **40**, 38.
10. Knight, S. A., *Chem. Ind.* (1967) 1920.

11. Schweighardt, F. K., Friedel, R. A., Retcofsky, H. L., *Appl. Spectrosc.* (1976) **30**, 291.
12. Taylor, S. R., Galya, L. G., Brown, B. J., Li, N. C., *Spectrosc. Lett.* (1976) **9**, 733.
13. Ouchi, K., *Carbon* (1966) **4**, 59.
14. Nandi, B. N., Belinko, K., Pruden, B. B., Denis, J. M., *Am. Chem. Soc., Div. Pet. Chem., Prepr.* (1977) **22**(2), 733.

RECEIVED August 5, 1977.

6

New Techniques for Measuring PNA in the Workplace

R. B. GAMMAGE, T. VO–DINH, A. R. HAWTHORNE,
J. H. THORNGATE, and W. W. PARKINSON

Health and Safety Research Division, Oak Ridge National Laboratory,
Oak Ridge, TN 37830

A gap exists between the crude techniques available for measuring PNA compounds in the workplace and the sophisticated analytical tools used in the laboratory for a much fuller characterization of pollutants from synfuel operations such as tar sand and oil shale processing. Real-time or near-real instruments suitable for use by industrial hygienists are needed urgently to measure fugitive emissions. Several new instruments and instrumental techniques are described that could satisfy some of these needs. They include second-derivative UV absorption, synchronous luminescence, and room-temperature phosphorescence spectrometries, a portable mass spectrometer, differential sublimation, and thermoluminescence. Already, studies to evaluate the practicality of these approaches have indicated a suitability for monitoring naphthalene and its alkyl derivatives at parts-per-billion (ppb) concentrations either in the vapor or the solution phase, trace amounts of phenolic compounds in by-product water, and for the rapid analysis of samples filtered or spotted on paper adsorbents.

The processing of fossil fuels such as tar sand, oil shale, and coal in synthetic fuel cycles produces a wide range of noxious chemicals. Some of the most difficult to handle in providing adequate occupational protection are the polynuclear aromatic (PNA) compounds that have a potential for carcinogenic action as initiators, cocarcinogens, or tumor promoters. Often the dangers are compounded by pronounced synergistic effects. This chapter confines itself primarily to the instruments and techniques needed for occupational monitoring for synfuel pollutants of polycyclic compounds with two or more benzene rings.

At present the industry is burdened only lightly with federal regulations for occupational exposure to multi-ring PNA. There is but a single standard; the Occupational Safety and Health Administration (OSHA) has issued an ordinance limiting emissions (from coke ovens) that is to be phased in over a period of three years—150 μg of benzene-soluble particulates per cu m of air, averaged over 8 hr. The benzene-soluble fraction of the particulates, otherwise known as coal-tar-pitch volatiles (CTPV), contain PNA compounds of which some, like 3,4-benzopyrene (BaP), are carcinogens. The Standards Advisory Committee on Coke-Oven Emissions has recommended recently the adoption of BaP as an indicator PNA carcinogen to be present in particulate matter at concentrations no greater than 0.2 μg/m^3 (1). Additional standards can be expected to follow the appearance of practical monitoring techniques suitable for industrial hygiene usage, and clearer identification of offending individual chemicals or groups of compounds. The latter goal is being approached through testing for mutagenicity (Ames Test (2)) of chemically fractionated components of process streams (3). The magnitude of the occupational health hazards also will be appreciated better by making more detailed retrospective, as well as prospective, epidemiological studies.

In order to ensure that occupational risks are of an acceptable magnitude, a fairly wide range of monitoring devices will be needed. Since what constitutes a safe exposure level is not known, it would seem prudent to make initial monitoring programs as broad-based as possible. Vapors, airborne particles of respiratory size, and surfaces contaminated with tar and pitches are pathways of occupational exposure. Table I serves to highlight the gap that exists between the rather crude and chemically nonspecific tools available to the industrial hygienist and the sophisticated, expensive techniques available to the researcher for the often ultrasensitive analysis of specific compounds. It also includes our estimate of what is required to bridge this gap.

A specific example will illustrate the type of difficulties currently being faced in monitoring for occupational protection. Inhaled particulates that contain PNA present a potential risk via the induction of lung cancer (4). The presently used method for in-plant monitoring involves the collection of particulates with glass fiber and silver membrane filters, extraction with hot benzene, and a determination of the weight of benzene-soluble material by the difference between the filter weights before and after extraction (5). This method is easy to follow, requires little sophisticated or expensive equipment, and can be operated by relatively inexperienced technicians with a minimum of supervision. The draw-

Table I

Crude PNA Monitoring Methods Available to Industrial Hygienists

1. Benzene solubles within particulate matter.
2. Gross emission from UV irradiation of the skin, clothing or inanimate objects.

Highly Sensitive and Selective Research Tools for PNA Characterization

1. Gas Chromatography (or High Pressure Liquid Chromatography) —Mass Spectroscopy.
2. Matrix Isolation, Fourier Transform—Infrared.
3. Chemical Separation and Gas Liquid Chromatography.
4. X-ray Excited Optical Luminescence.
5. Laser Induced Fluorescence.

Current Needs for Pilot Plant Monitoring to Bridge the Gap

Techniques and Instruments which are:

1. Selective for specific compounds or groups of compounds.
2. Sensitive at OSHA concentration limits.
3. Real-time or near real-time in producing analytical data.
4. Easy to operate by technicians (fitted with microprocessors).
5. Portable or semi-portable for personnel monitoring or leak detection.
6. Reasonable cost.

backs, however, are quite numerous. Apart from being relatively insensitive (0.1-mg benzene-soluble material) and nonspecific, one is making a measurement over an extended time of worker exposure and then one is required to wait some considerable time before the analytical result becomes available. The difficulty would be increased if, in addition, the BaP within the CTPV (*1*) would need to be analyzed, since more complex and time consuming chemical separations are required then.

A real-time monitor for PNA compounds, or for an indicator such as BaP, is highly desirable. Evans (*6*) has suggested that alternatives to the benzene-soluble fraction of total particulate matter, as a measure of biologically active carcinogenic material, should be explored; he outlines two techniques that might prove superior to present methods based on benzene solubles. One is a real-time BaP monitor for particulates collected on a moving filter tape that uses laser-excited fluorescence at low temperature. The other involves selective solvation of collected particulate matter followed by a total fluorescence scan of the type that Baird–Atomic has developed for the U.S. Coast Guard to monitor samples of oil in water. We are exploring the potentials of the room-temperature phosphorescence technique for real-time or near real-time monitoring as will be described in a later section.

Our program to develop new instruments suitable for use by industrial hygienists in the synthetic fuels industry is still in the fledgling stage. This chapter aims at acquainting the reader with our thoughts and earliest attempts to improve and increase in number the monitoring devices that are available to help in protecting workers from gases such as CO, H_2S, CS_2, NH_3, HCN, and phenols and high-boiling aromatic compounds in converted fossil fuels such as tar sand, oil shale, and coal.

At sites such as the University of Minnesota, Duluth, the Land O'Lakes, Peron, Minnesota, and Pike County, Kentucky, low-Btu coal gasifiers based on proven technology are reappearing in the USA. Second generation coal converters (e.g., HYGAS, Chicago and H-Coal, Catlettsburg, Kentucky) are in advanced pilot-plant stages of development. To prove within a few years that these technologies are acceptable environmentally requires the rapid development of a new generation of monitoring devices. The urgency of this task permits an element of prematurity in the reporting of several of the approaches that we are following. Proving that any of these analytical techniques will be of real value in the workplace lies in the future.

Second-Derivative UV Absorption Spectrometry

This instrument has been available commercially for a few years (Lear Siegler Model SM400) and offers strong potential for monitoring the lower-boiling PNA and other simple aromatic molecules. Thus far it has been sold primarily for the in situ monitoring of NO_x and SO_2 inside smoke stacks. Its attractiveness for us is in its ability to provide rapid analysis of synfuel pollutants.

The second-derivative spectrometer produces a signal which is proportional to the curvature of the absorption spectrum with respect to wavelength. This signal is proportional to the concentration of the measured compound and is additive for the various component spectra resulting from a mixture of compounds. The curvature is often quite large and specific to individual compounds, thus allowing selective analysis of component compounds in rather complex samples. Many of the components found in the process streams of fossil–fuel conversion facilities have quite distinctive second-derivative spectra. For details of the second-derivative spectrometer and of the method of data analysis used for a mixture of compounds, the reader should consult References 7, 8, 9, and 10.

The spectrometer is equipped with a 1-m multipass (up to a 32-m path length) sample cell for direct monitoring of vapor-phase pollutants. It also is capable of analyzing liquid samples using a 1-cm quartz cell. In addition to the scanning mode of operation, where a complete spectrum is obtained, the spectrometer can be operated in a dwell mode at a specific wavelength, allowing for real-time monitoring of a selected compound.

This instrument can be used for a variety of industrial hygiene and environmental monitoring applications. The vapor-phase mode of analysis may provide for area monitoring of fugitive emissions or for leak detection when trying to locate point source releases. It also may be useful as a detector for vapors thermally desorbed from adsorbants or swipe samples. The liquid mode of analysis will prove valuable in the direct monitoring of aqueous streams. Particulate organic matter collected on filter papers or adsorbed organic vapors collected in sampling tubes (charcoal, Tenax, etc.) can be dissolved into a suitable solvent and the spectrometer can serve as the reader for a dosimetry system. Two specific examples involving pollutants of practical concern will serve to demonstrate the usefulness and versatility of the technique.

The first example involves the vapors of naphthalene and its alkyl derivatives. These volatile compounds are ubiquitous to fossil–fuel operations. Only naphthalene itself has an OSHA concentration limit, which is 10 ppm. The most abundant member of the naphthalene family in, for example, the condensate from the Synthane gasification process (11) is 2-methylnaphthalene, melting at 35°C. Although it is recognized as a fairly active tumor promoter (12), little consideration has been given to the effects upon health of chronic exposure to its vapor. There is no OSHA limit on the maximum permissible concentration.

Figure 1 shows portions of second-derivative spectra of naphthalene vapor at 90 ppb, of 2-methylnaphthalene vapor at 73 ppb, and of a mixture containing 90 ppb naphthalene and 73 ppb 2-methylnaphthalene. Note the almost complete separation of the two peaks even though they are only about 3 nm apart. Using the least-square fitting procedure described in References 7 and 10, an analysis of 86 ± 2 ppb (error quoted is 1σ for the least-squares fit) for naphthalene and 71 ± 2 ppb for 2-methylnaphthalene was obtained. The selectivity in this example is sufficiently good so that graphical analysis also can be used. The linear relationship between instrument response and known concentration is shown in Figure 2 with the analytical curve for 2-methylnaphthalene.

The second example involves the measurement of aqueous pollutants. Here again 2-methylnaphthalene is very prominent among the PNA compounds and was found to be the most abundant PNA in the by-product water of oil shale retorting and Synthane coal gasification (11). Again, the second-derivative spectrometer could serve as a convenient analytical tool. Phenol bearing wastewater, however, presents an apparently greater problem on account of the greater solubility of the acidic phenolic compounds. Figure 3 is a spectrum of phenol in water. The

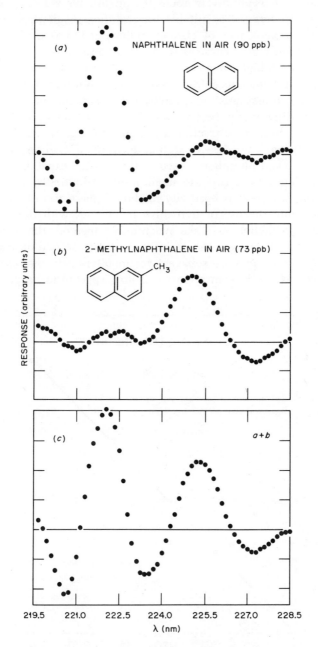

Figure 1. Second-derivative spectra of naphthalene and 2-methylnaphthalene in air

lower limit of detectability is about 0.1 μg/mL for a 1-cm quartz cell. This mode of operation is suitable for real-time monitoring of aqueous streams at various stages of cleanup following the synfuel operations. The analytical response curve is shown in Figure 4. Note the departure from linearity at high concentrations. A method of retaining a linear relationship at high concentrations has been developed (7). Further examples of complex mixture analysis may be found in Reference 10. Improved sensitivity may be obtained by using a longer path length and one of various chemical extraction techniques to concentrate the solute.

The principal advantage of second-derivative UV absorption relative to fluorescence and phosphorescence methods is the capability of using a long path length for gas-phase monitoring. Admittedly, luminescence spectroscopy has greater sensitivity for liquid-phase monitoring, and we are investigating the use of derivative techniques with these methods, perhaps in conjunction with the synchronous method, thus making the peaks sharper. However, the sensitivity of the second-derivative UV absorption method may be sufficient for monitoring in the occupational environment where concentration levels are expected to be greater than

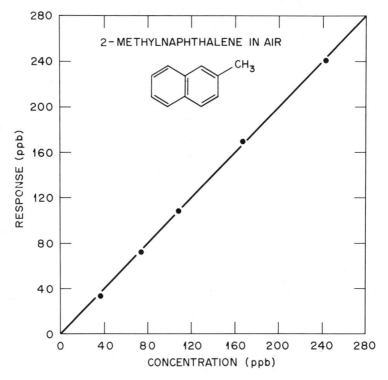

Figure 2.   Analytical curve for 2-methylnapthalene in air

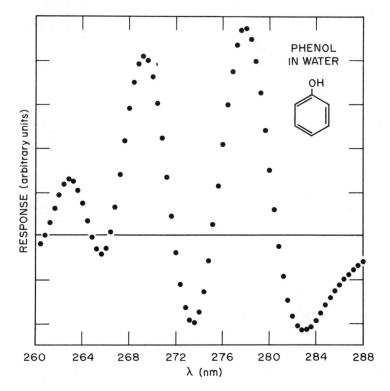

Figure 3. Second-derivative spectrum of phenol in water

in the general environment. Previously reported results showing sensitivity to the 5-ppb level for naphthalene in air and 1 ng/mL for anthracene in solution indicate that this technique may indeed be sufficiently sensitive for many applications (*10*).

Future plans call for the incorporation of a microprocessor into the spectrometer to provide instrument control and to act as an interface with a microcomputer that can be used for data storage and analysis. The objective is the development of portable, real-time instruments, such as a monitor for measuring phenolic compounds and/or alkyl naphthalenes in wastewater.

Synchronous Luminescence Spectrometry

Luminescence spectroscopy, with its excellent sensitivity for trace analysis, has limited applicability for analysis of complex mixtures of organic molecules. The most successful techniques involve cryogenic temperatures with matrix isolation or nanosecond time resolving capability (*13*). While being highly selective techniques, they are not easily

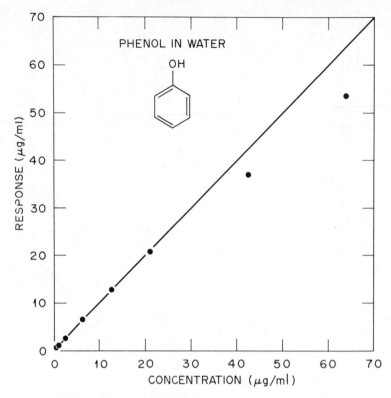

Figure 4.   Analytical curve for phenol in water

applicable to routine industrial hygiene. A new method has been developed lately (*14, 15*) that is based on the so-called synchronous luminescence technique (*16*). Often used in identifying crude oils (*16, 17*), this technique has only been applied in the past in an empirical manner.

Commercial spectrometers, such as the Perkin–Elmer MPF–43A fluorescence spectrometer, that allow interlocking of excitation and emission monochromators lately have become available for utilizing this underexploited analytical technique. The synchronous luminescence technique reduces the complexity of the luminescence spectrum of a compound compared with a conventionally obtained luminescence spectrum. One can, therefore, better tackle the analysis of fairly complex mixtures without resorting to techniques that are expensive or excessively time consuming.

It was suggested first by Lloyd (*15, 16*) that one can vary simultaneously the excitation and emission wavelengths with a fixed frequency difference $\Delta\lambda$, between the two. The luminescence is given by (*14*):

$$I_L(\lambda,\lambda') = k \cdot c \cdot E_X(\lambda') \cdot E_M(\lambda)$$

with $\lambda' = \lambda - \Delta\lambda$, where I_L is the observed luminescence intensity, k is a constant, c is the concentration of the analyte, $E_X(\lambda')$ is the excitation spectrum at fixed emission wavelength, $E_M(\lambda)$ is the emission spectrum at a fixed excitation wavelength, λ' is the excitation monochromator wavelength, and λ is the emission monochromator wavelength. The

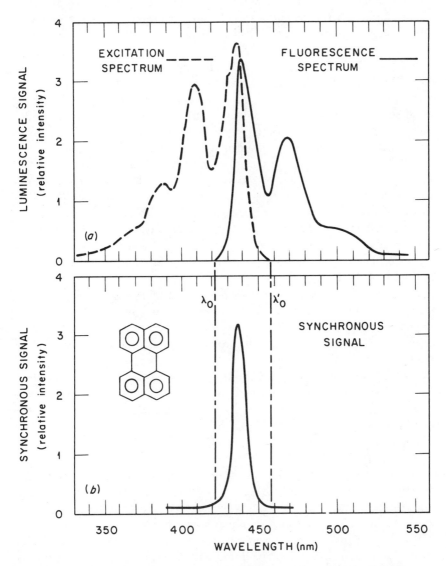

Figure 5. Fluorescence spectra of perylene in ethanol. (a) Fluorescence spectrum (——) at fixed excitation wavelength ($\lambda_{ex} = 407$ nm); excitation spectrum (– – –) at fixed emission wavelength ($\lambda_{em} = 470$ nm). (b) Synchronous fluorescence signal ($\Delta\lambda = 3$ nm).

luminescence intensity in synchronous spectroscopy, as opposed to conventional spectroscopy, depends explicitly on the emission spectrum $E_M(\lambda)$ as well as on the excitation spectrum $E_X(\lambda')$ and the difference in wavelength, $\Delta\lambda$, which can be selected to improve the selectivity.

The maximum possible spectral simplification that one can achieve and the consequent improvement in selectivity is illustrated in Figure 5. Figure 5a shows typical fluorescence and excitation spectra of perylene. The fluorescence intensity will change if the excitation wavelength is varied; the basic intensity distribution, however, will not change. By choosing a 3-nm value for $\Delta\lambda$, so as to match the Stokes shift between the O–O bands in the excitation and emission spectra, the single emission peak (Figure 5b) is obtained in the synchronous spectrum. Here we have selected the strongest emission peak with the smallest half-width and caused the corresponding synchronous signal to be still narrower without a sizeable loss in the intensity at the peak maximum (14).

The ability to limit the emission spectrum within a certain spectral interval is the strong point of the synchronous technique when it comes to analyzing a mixture of compounds. An ethanol mixture of five PNA compounds with different ring sizes produces the conventional fluorescence spectrum (excitation wavelength 258 nm) reproduced in Figure 6a. The corresponding synchronous spectrum in Figure 6b, using a 3-nm wavelength interval, shows the striking simplification and improvement in resolution that is possible.

The family of linear ring PNA compounds shows synchronous emission for the O–O transition that increases to longer wavelengths with increasing ring number. This behavior could provide the means for a rapid screening type of analysis to provide information on the abundances of specific types of compounds.

An example of a practical application might be the search for PNA compounds in by-product water and more specifically for the naphthalene family of compounds discussed earlier. To examine a specific case, the conventional fluorescence spectrum of a dilute solution of the Synthane gasifier, by-product water in methanol is given in Figure 7a. This particular spectrum is produced at an excitation wavelength of 290 nm but does not change drastically at other excitation frequencies with the emission bands remaining diffuse and without much structure. Synchronous luminescence using a 3-nm wavelength interval produces the much more structured fluorescence emission spectrum shown in Figure 7b. It will be noted that a peak occurs at 327 nm in the region expected for naphthalene and its alkyl derivatives. Figure 8 shows the corresponding synchronous spectra for the single compounds of naphthalene, 1-methylnaphthalene, and 2-methylnaphthalene. The by-product water is, therefore, revealing an abundance of 2-methylnaphthalene, which is expected

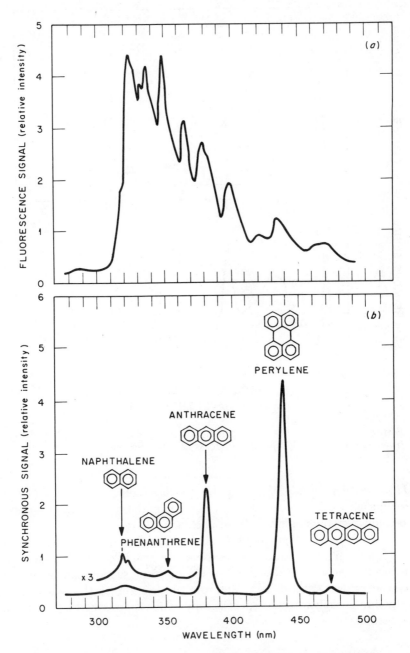

Figure 6. (a) Fluorescence spectrum of a mixture of naphthalene, phenanthrene, anthracene, perylene, and tetracene ($\lambda_{ex} = 258$ nm). (b) Synchronous signal ($\Delta\lambda = 3$ nm) of the same mixture.

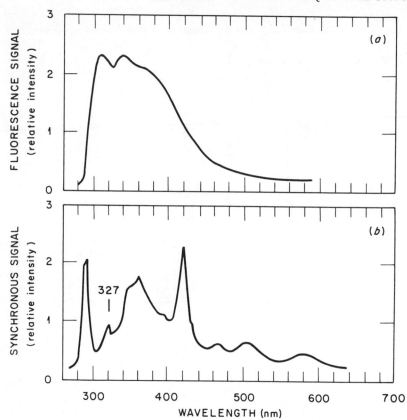

Figure 7. (a) Fluorescence spectrum of a solution of by-product water from Synthane coal gasification process ($\lambda_{ex} = 290$ nm). (b) Synchronous fluorescence ($\Delta\lambda = 3$ nm).

(11). Actual analysis of this water by gas–liquid chromatography (GLC) profiling gave 160, 32, and 1300 μg/L of naphthalene, 1-methylnaphthalene, and 2-methylnaphthalene, respectively (11).

A wide range of possibilities exist for further improvements in the near future. Two of these are the use of a second-derivative attachment so that the techniques described in the previous section can be applied to synchronous spectroscopy, and the application of the synchronous method to phosphorescence spectroscopy which is described in the following section.

Room-Temperature Phosphorimetry (RTP)

In conventional phosphorimetry, analysis is performed at cryogenic temperatures using rigid matrices in order to minimize collisional triplet quenching. At room temperature, rigidity can be achieved by adsorption of the sample onto a matrix-like backing such as filter paper. The adsorbate, if immobilized sufficiently by strong adsorbent–adsorbate forces, can be induced to phosphoresce strongly. The RTP has been observed from a number of salts of organic compounds adsorbed on a variety of supports such as silica, alumina, filter paper, asbestos, and cellulose (*18*). Further developments to transform this effect into an analytical tool have been reported by Winefordner et al. (*19, 20, 21, 22, 23*).

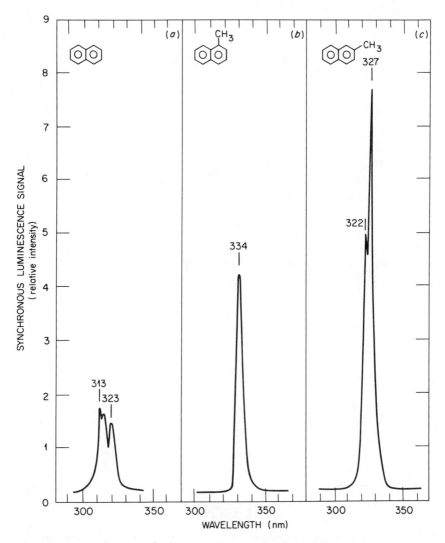

Figure 8.   Synchronous fluorescence (Δλ = 3nm) of: (a) naphthalene; (b) 1-methylnaphthalene; and (c) 2-methylnaphthalene

An RTP assay consists generally of four steps that are shown schematically in Figure 9. The purpose of the heavy-atom perturber is to encourage population of the triplet state via spin orbit coupling and intersystem crossing. The use of silver ion, as silver nitrate, is very effective in bringing this about and thereby enhancing the phosphorescence emission of many adsorbed PNA compounds.

The experimental details have been described already in detail (19, 20, 21). The paper substrate can be selected, or pretreated, in a way designed to optimize the phosphorescent intensity. Sample delivery consists of spotting 3 μL of sample solution on a circle of filter paper 0.6 cm in diameter. Predrying with an IR heating lamp proved to be more efficient than the use of desiccators, blowers, or ovens (19, 20, 21, 22). Maintenance of the dried state of the material during spectroscopic measurement is achieved by blowing warm, dried air through the sample compartment. The previously mentioned Perkin Elmer spectrofluori-

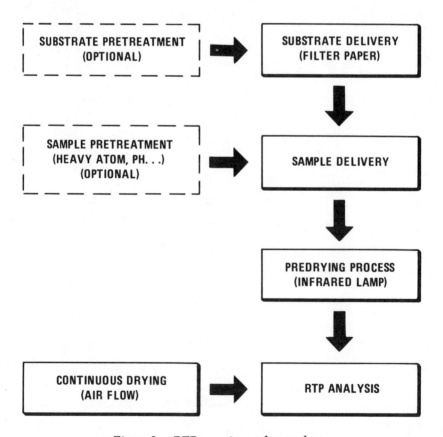

Figure 9.   RTP experimental procedure

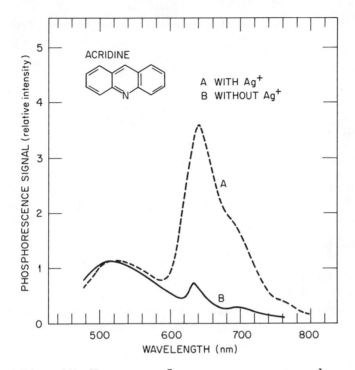

*Figure 10.   Heavy-atom effect on room-temperature phos-
phorescence of acridine: (a) with silver nitrate (0.25M),
52-ng sample; (b) without silver nitrate, 210-ng sample.*

meter, with a phosphoroscope attachment, was used to obtain the RTP
data presented here.

The limits of detection of particular compounds were determined
using a large monochromator bandpass (15 nm). The adsorbent support
was Schleicher and Schuell 591–C filter paper. All solutions were made
up in an ethanol–water mixture (3 to 1 by volumes). The silver nitrate
solution was 0.25M. An example of enhancement of phosphorescence by
the heavy-atom effect is shown in Figure 10. The emission peak at
~ 640 nm is characteristic of acridine. The marked intensification caused
by the silver ion is quite clear. The lower limit of detectability can be
lowered an order of magnitude for several PNA compounds; relevant
data are contained in Table II. The limit of detection is defined as the
amount of analyte resulting in a signal-to-noise ratio of 3. Of the eight
compounds investigated, the lower limit of detectability is in the subnanogram range, except for the 2-methylnaphthalene.

There are several ways in which the selectivity and sensitivity can yet
be improved. These include techniques of phase and time resolution,
derivatization of the emission bands, and synchronization of excitation

Table II.   Limit of Optical Detection (L.O.D.) by RTP Technique

Compound	Excitation Wavelength λ_{ex} [nm]	Emission Wavelength λ_{em} [nm]	Absolute L.O.D. [ng] With AgNO$_3$	Absolute L.O.D. [ng] Without AgNO$_3$
Acridine	359	643	0.02	—
Chrysene	278	515	0.013	0.27
1,2,3,4-Dibenzanthracene	290	570	0.05	—
Fluoranthene	360	570	0.05	—
Fluorene	278	440	0.25	2
Phenanthrene	300	567	0.06	0.22
Pyrene	346	599	0.15	0.35

and phosphorescence signals; and each will be investigated in the near future. This simple and novel method of analysis opens up a host of possibilities for monitoring organic pollutants at trace levels. One of these possibilities with RTP is perhaps that PNA filtered from the air onto a moving paper tape can be engineered to provide a near real time monitor of both vapors and particulates within a worker's breathing zone.

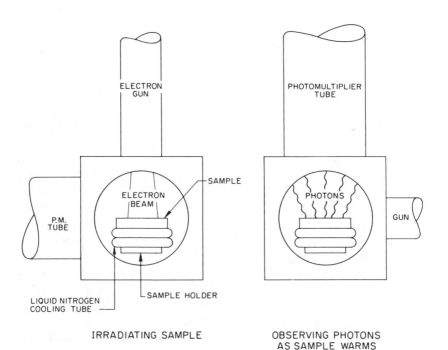

Figure 11.   Schematic of the operation of the RTL unit

Thermoluminescence Analysis

Many substances emit light by radiothermoluminescence (RTL) if they are exposed to ionizing radiation and then warmed. Most studies of this phenomenon involve inorganic materials, although organic materials such as amino acids and enzymes (*24, 25*) also exhibit RTL. Thermoluminescence has been observed recently in PNA compounds (*26*). This latter work has prompted the investigation of RTL for possible use in health protection measurements. For most applications, samples collected in the work area would be placed in a solvent before introducing them to the RTL unit.

The use of RTL as an analytical tool differs from its more conventional use in radiation dosimetry; the samples must be exposed to very high doses (1000 Rad) of radiation and the irradiations generally must be made at 80°K, or below. These requirements preclude using existing radiation-dosimeter readout systems. A new apparatus has been built (Figure 11), for measurements of RTL from PNA compounds in which a small sample is put into a vacuum chamber and cooled to ca. 80°K using liquid nitrogen. After cooling, the chamber is evacuated and the sample is irradiated with low-energy electrons (≤ 1 keV). Electrons were chosen for the irradiations because high doses can be delivered to the samples in reasonably short times with readily available beam currents. Little external radiation is produced by the electron beam but its use does necessitate a vacuum system. Without removing the sample from the chamber, the electron source is replaced with a photomultiplier tube and, as the sample warms to room temperature, the RTL is measured. More detail and possibly greater selectivity could be obtained by measuring the spectrum of the emitted light. There is evidence that the more biologically active materials have the largest RTL responses, which would add to the attractiveness of this analytical method for PNA. The technique also could be valuable for rapid screening of samples based on gross measurement of the RTL. Although inexpensive, the liquid nitrogen required may not be readily available, which could be a drawback to this technique.

Separation by Differential Sublimation

Differential sublimation can separate and purify complex mixtures of organic materials. This was shown in a previous program where it was used to purify parasexiphenyl for use as a scintillator (*27*). A closely related technique was used elsewhere to separate PNA's for subsequent analysis by fluorescence spectroscopy (*28*). A system has been built (Figure 12), to investigate the applicability of this technique to the rapid evaluation of samples taken at coal conversion facilities.

The sample is placed in a furnace where hot gas flows over and vaporizes it. Recrystallization occurs in a tube carrying a temperature gradient along its length. The apparatus has been designed for maximum flexibility and consists of heaters for the carrier gas and the sample, and a furnace designed to provide a linear temperature gradient along a quartz tube. The temperatures of the carrier gas, the sample, and the inlet and outlet temperatures of the gradient furnace can all be adjusted independently. Recrystallization occurs in the quartz tube. Using quartz will

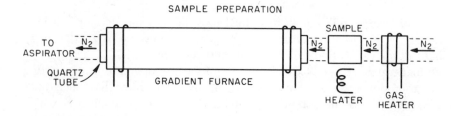

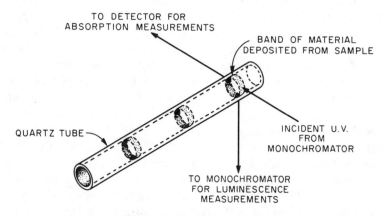

Figure 12.   Diagram of the differential sublimation apparatus

simplify subsequent analysis of the separated samples by UV-absorption spectroscopy. To prevent damage from excessive pressures, an aspirator is used to pump the carrier gas through the system. This also disposes of unwanted materials that recrystallize at temperatures below those at which the furnace is set.

Hopefully, this procedure will provide rapid and simple separations of complex samples so that the selectivity required in subsequent analysis by UV-absorption or luminescence emission can be decreased. It also may be possible to use an apparatus of this type for continuous air monitoring in work areas. With additional instrumentation, it could provide continuous monitoring of several substances in real time. The apparatus also can purify laboratory samples of the important PNA's to provide better standards for other detection and measurement techniques.

Portable Mass Spectrometer

Some of the most detailed characterization studies performed on component streams in conversion processes have used a gas chromatograph–mass spectrometer (GC–MS) as the analytical tool. These instruments are well established as laboratory analytical tools of high sensitivity and reproducibility. The use of such systems as part of a monitoring program for PNA exposures would be prohibitively expensive.

A portable, quadrupole mass spectrometer, however, has been developed to monitor organic vapors (29). Because this instrument has proven valuable in the U.S. Army industrial hygiene program, the design of a second generation instrument with the capability of attaching a portable gas chromatograph is proceeding. A microcomputer-controlled, portable, quadrupole mass spectrometer is being developed also as an ambient air monitor under an interagency agreement between NASA, EPA, ERDA, and various branches of the Department of Defense. This instrument, although small and relatively inexpensive, will be capable of making rapid, repetitive measurements much faster than a GC–MS while making the appropriate matrix calculations to correct for spectral interferences.

One of the instruments obtained by ERDA will be evaluated at ORNL to assess its potential as a monitor for hazardous by-products from alternative energy sources. Investigations to determine the desirability of a membrane inlet system for concentrating organic vapors are planned. The feasibility of using a portable gas chromatograph with the portable mass spectrometer, when a complex mixture analysis is required, is also being studied.

Literature Cited

1. *Occupational Safety and Health Reporter* (1975) 4, 1639.
2. Ames, B. N., Kammen, H. O., Yamasaki, E., *Proc. Nat. Acad. Sci., U.S.A.* (1975) 72, 2423.
3. Rubin, I. R., Guerin, M. R., Hardigree, A. A., Epler, J. L., *Environ. Res.* (1976) 12, 358.
4. Mazumdar, S., Redmond, C., Sollecito, W., Sussman, N., *J. Air Pollut. Control Assoc.* (1975) 25(4), 388.
5. "Report on Analytical Methods Used in a Coke Oven Effluent Study," *HEW Pub.* No. (NIOSH) 74-105, 1974.

6. Evans, J. M., "Preliminary Recommended Guidelines for Coal Gasification Units," Enviro Control, Inc., Rockville, Maryland, 1977.
7. Hawthorne, A. R., Thorngate, J. H., "Improved Analysis from Second-Derivative Spectrometry," *Appl. Opt.* (1978) **17**(5), 724.
8. Williams, D. T., Hager, R. N., Jr., *Appl. Opt.* (1970) **9**(7), 1597.
9. Hager, R. N., Jr., Anderson, R. C., *J. Opt. Soc. Am.* (1970) **60**(11), 1444.
10. Hawthorne, A. R., Thorngate, J. H., "Application of Second-Derivative UV-Absorption Spectrometry to PNA Analysis," *Appl. Spectro.* (1978) in press.
11. Ho, C. H., Clark, B. R., *Coal Technology Program Annual Interim Report for Fiscal Year Ending June 30, 1976,* ORNL-4208, p. 95.
12. Schmeltz, I., Hoffman, D., Wynder, E. L., "Trace Substances in Environmental Health VIII," D. D. Hemphill, Ed., p. 281, University of Missouri (Columbia), 1974.
13. Wehry, E. L., Mamantov, G., "Abstracts of the First ORNL Workshop on Polycyclic Aromatic Hydrocarbons: Characterization and Measurement with a View Toward Personnel Protection," (1976) ORNL/TM-5598, 43.
14. Vo-Dinh, T., *Anal. Chem.* (1978) **50**, 396.
15. Vo-Dinh, T., Gammage, R. B., Thorngate, J. H., Hawthorne, A. R., *Environ. Sci. Technol.* (1978), in press.
16. Lloyd, J. B. F., *J. Forensic Sci. Soc.* (1971) **11**, 83, 153, 235.
17. Philip, J., Souter, I., *Anal. Chem.* (1976) **48**, 520.
18. Schulman, E. M., Walling, C., *Science* (1972) **178**, 53.
19. Wellons, S. L., Paynter, R. A., Winefordner, J. C., *Spectrochim. Acta* (1974) **30A**, 2133.
20. Paynter, R. A., Wellons, S. L., Winefordner, J. D., *Anal. Chem.* (1974) **46**, 736.
21. Vo-Dinh, T., Lueyen, E., Winefordner, J. D., *Anal. Chem.* (1976) **48**, 1186.
22. Vo-Dinh, T., Lueyen, E., Winefordner, J. D., *Talanta* (1977) **24**, 146.
23. Vo-Dinh, T., Walden, G., Winefordner, J. D., *Anal. Chem.* (1977) **49**, 1126.
24. Carter, J. G., Nelson, D. R., Augenstein, L. G., *Arch. Biochem. Biophys.* (1965) **111**, 270.
25. Nelson, D. R., Carter, J. G., Birkhoff, R. D., Hamm, R. N., *Radiat. Res.* (1967) **32**, 723.
26. D'Silva, A. P., Oestreich, G. J., Fassel, V. A., *Anal. Chem.* (1976) **48**, 915.
27. Perdue, P. T., Arakawa, E. T., *Nucl. Instrum. Methods* (1975) **128**, 201.
28. Monkman, J. L., Dubois, L., Baker, C. J., *Pure Appl. Chem.* (1970) **24**, 731.
29. Evans, J. E., Arnold, J. T., *Environ. Sci. Technol.* (1975) **9**, 1134.

RECEIVED August 5, 1977. Research sponsored by the Energy Research and Development Administration under contract with the Union Carbide Corporation.

Characterization of Mixtures of Polycyclic Aromatic Hydrocarbons by Liquid Chromatography and Matrix Isolation Spectroscopy

G. MAMANTOV, E. L. WEHRY, R. R. KEMMERER,
R. C. STROUPE, and E. R. HINTON

Department of Chemistry, The University of Tennessee,
Knoxville, TN 37916

G. GOLDSTEIN

Analytical Chemistry Division, Oak Ridge National Laboratory,
Oak Ridge, TN 37830

Mixtures of polycyclic aromatic hydrocarbons (PAH) as found in naturally occurring oils and petroleum, or synthetic oils produced by coal conversion processes, are too complex for characterization by any single technique. The approach taken frequently is to apply preliminary separation procedures of various kinds, including distillation, extraction, and chromatographic methods, then attempt to analyze the separated fractions by a high resolution technique. In this chapter we describe the liquid chromatographic separation of mixtures of PAH using poly(vinyl pyrrolidone) (PVP) as the stationary phase and isopropyl alcohol as eluant. The resulting fractions are examined by matrix isolation (MI) Fourier transform infrared (FTIR) and fluorescence spectroscopy.

Mixtures of polycyclic aromatic hydrocarbons (PAH), as encountered in naturally occurring oils and petroleum, or in synthetic liquid fuels produced by coal liquefaction or oil shale processing, are far too complex for total characterization by any single analytical technique. The approach taken usually in the characterization of such materials is to apply pre-

0-8412-0395-4/78/33-170-099$05.00/1

liminary separation procedures of various kinds (including distillation, extraction, and chromatographic methods), followed by examination of the individual fractions by a high-resolution spectroscopic technique. The most obvious example of such an approach is the combination of gas chromatography and mass spectrometry (GC–MS), which has been used widely in the characterization of the organic constituents of liquid fuels (1, 2, 3, 4, 5). While GC–MS has proven to be an exceedingly powerful tool for the identification of specific compounds in complex mixtures, the technique is not totally devoid of shortcomings. Among these are: (a) difficulty in distinguishing between isomeric compounds (6), unless they have been separated on the GC column, which is not necessarily easy to accomplish (7); (b) difficulty in achieving rapid analyses of complex samples; and (c) difficulty in achieving high quantitative accuracy and precision (8). We are engaged currently in a study of the use of liquid chromatography (LC), followed by either fluorescence or IR matrix isolation spectrometric examination, as a technique for the identification and quantification of PAH in coal liquids and shale oils.

A detailed description of the LC method used in this work has appeared previously (9). The stationary phase in the LC procedure is cross-linked poly(vinyl pyrrolidone) (PVP); a polar solvent (usually isopropyl alcohol) is used as the eluant. The procedure resolves PAH in complex samples into groups of compounds determined primarily by the number of aromatic rings; resolution of complex samples into individual pure compounds cannot be achieved normally in a realistic time period. Spectroscopic techniques to be used in the analysis of these LC fractions accordingly must exhibit high resolution, be capable of detecting closely related (including isomeric) compounds in small quantities, and be capable of providing quantitative information pertaining to the quantities of specific compounds present in a given fraction. These requirements are exceedingly stringent, and there is probably no spectroscopic procedure extant which can fully satisfy them.

Fluorescence spectroscopy has been used widely in the analysis of PAH (10) because of its very high sensitivity; when state of the art instrumentation is used, fluorescence methods are capable of detecting less than 10^{-12} g of intensely fluorescent analytes in liquid solution (11). Unfortunately, fluorescence in fluid media is generally not useful for quantitative analyses of individual fluorophores in mixtures for two reasons. First, fluorescence spectra of most organic compounds in fluid media are broad and relatively devoid of fine structure; hence, the fluorescence spectrum of a mixture of fluorophores tends to be an intractable mass of overlapping bands. Second, and more important, interference effects (quenching and energy transfer, excited dimer formation, and inner filter effects) are othen important in the fluorimetric analysis of a mixture of compounds in liquid solution (12). In effect, therefore, fluorescence in liquid solution is highly susceptible to matrix effects, wherein the fluorescence intensity observed for a particular compound depends not only upon the concentration of that compound, but upon the concentrations of other sample constituents. Under these circumstances, obtaining accurate quantitative analyses for individual compounds in mixtures is a virtual impossibility; thus, despite its very high sensitivity, fluorescence spectrometry has fallen into some disrepute as a practical quantitative analytical technique for the characterization of coal liquids, shale oils, and similar samples.

Inasmuch as quenching and energy transfer phenomena often proceed by diffusional mechanisms, it is obvious that the only practical approach to their suppression is to perform fluorescence measurements in rigid media. The conventional approach to performing fluorescence measurements in rigid media is to freeze a liquid solution of the sample, usually at liquid nitrogen temperature (77°K). This approach leads to a considerable increase in fine structure; in some specific cases, the solute can be made to occupy specific sites in a quasicrystalline frozen solution, in which case very sharp, line-like fluorescence can be observed from PAH (the Shpol'skii effect (13)). Application of the Shpol'skii effect to both qualitative and quantitative analyses of PAH in mixtures has been described (14). The frozen-solution procedure appears highly valuable as a technique for fingerprinting specific PAH present in very small quantities in complex mixtures, but (for reasons detailed elsewhere (15, 16, 17)) it has important limitations as a procedure for quantitative analyses in mixtures. The principal obstacle to performance of accurate quantitative analyses, over a wide concentration range, by frozen solution fluorimetry is the tendency for the solutes to aggregate (occasionally in microcrystalline domains) as the solution is frozen. In order to realize fully the benefits of fluorescence spectrometry in rigid media for quantitative analysis, it appears necessary to form the rigid matrix by some process other than freezing of a liquid solution.

The approach to low-temperature spectroscopy that we have chosen to pursue is termed matrix isolation (MI). In MI, the sample is vaporized and then mixed with a large excess of a diluent gas (18, 19, 20). The gaseous mixture then is deposited on a window at low temperature for spectroscopic examination. The purpose of mixing the sample with the solvent (matrix gas) in the vapor phase is to secure an essentially random distribution of solute molecules, such that each analyte molecule has only matrix gas molecules as nearest neighbors. If this objective is achieved, and if the dilution is sufficiently great that the average distance between any two solute molecules is sufficiently large, then the fluorescence of any one analyte in a complex sample should be essentially unperturbed by the other constituents of the sample. In this way, it should be possible to perform fluorometric quantitative analyses in complex samples over wide concentration ranges.

While luminescence in vapor-deposited matrices accordingly should be a powerful technique for detection and quantitation of subnanogram quantities of PAH in complex samples, it suffers from two major limitations. First, it is obviously limited to the detection of molecules which fluoresce or phosphoresce, and a number of important constituents of liquid fuels (especially nitrogen heterocyclics) luminesce weakly, if at all. Second, the identification of a specific sample constituent by fluorescence (or phosphorescence) spectrometry is strictly an exercise in empirical peak matching of the unknown spectrum against standard fluorescence spectra of pure compounds in a library. It is virtually impossible to assign a structure to an unknown species a priori from its fluorescence spectrum; qualitative analysis by fluorometry depends upon the availability of a standard spectrum of every possible sample constituent of interest. Inasmuch as this latter condition cannot be satisfied (particularly in view of the paucity of standard samples of many important PAH), it is apparent that fluorescence spectrometry can seldom, if ever, provide a complete characterization of the polycyclic aromatic content of a complex sample.

For these reasons, we are engaged also in a study of the use of IR spectroscopy in the characterization of LC fractions of coal liquids and shale oils. IR spectroscopy has not been used previously in problems of this type, which requires high sensitivity and resolution, because it commonly has been assumed that IR was insufficiently sensitive and incapable of exhibiting adherence to Beer's law over wide concentration ranges (21). Indeed, IR spectrometry using conventional (dispersive) instrumentation and conventional sample-handling techniques (solutions, KBr discs, and mulls) is not likely to be broadly useful for detecting and quantitating PAH in liquid fuels. However, use of the MI technique described above, in conjunction with multiplexing interferometric Fourier transform (FT) instrumentation (22) should greatly improve the sensitivity and quantitative reliability of IR spectrometry (23, 24). The combination of IR and fluorescence should be an especially powerful one, because IR is a universal technique which provides spectral information related intimately to molecular structure; hence, it should be suitable for precisely those analytical problems in which the more sensitive fluorometric technique is incapable of providing useful results.

Experimental Section

Detailed descriptions of the LC procedure (9) (using PVP as stationary phase and isopropyl alcohol as eluant), and of the MI spectroscopic procedures for FTIR (*23, 24*) and fluorescence (*16*) measurements, have appeared previously. In the present work, a sample representing approximately 50 mg of an oil sample was dissolved in 0.25-mL isopropyl alcohol and separated on a 1.25- × 45-cm glass column packed with 75–100-μm particles of PVP (Polyclar AT, Serva Feinbiochemica). Eluant (isopropyl alcohol) flow rate was 1.2 mL/min, the temperature was 62°C, and system pressure was less than 20 psi. The course of the separation was followed by monitoring the UV absorbance of the eluant at 254 nm. Fractions were collected based either on the observed UV chromatogram or upon elution volumes determined previously with standard PAH samples.

The desired volume (typically 50–100 μL or more) of a particular fraction was added, by GC syringe or Eppendorf pipet, to the Knudsen cell (described in detail previously (*24*)); the solvent was removed by blowing a stream of air through the Knudsen cell at room temperature, after which the cell was evacuated and heated to the desired temperature for vacuum sublimation of the solutes. The vapor effusing from the Knudsen cell was mixed with a large excess (10^5:1–10^7:1 on a mole basis) of matrix (nitrogen), and the gaseous mixture was deposited on a window of CsI (for IR) or sapphire (for fluorescence) mounted in the evacuated head of a closed-cycle cryostat (*24*). Deposition temperatures were normally in the 15–18°K range. Quantitation of specific sample constituents was effected by successive standard additions.

Results and Discussion

Inasmuch as PAH tend to elute from PVP in order of increasing number of rings (*9*), it was convenient to collect seven fractions in the LC of coal liquids and shale oils. In general, these fractions represented monoaromatics (Fraction I), diaromatics (II), triaromatics (III), pericondensed four-ring compounds (IV), catacondensed four-ring compounds (V), pericondensed five-ring PAH (VI), and other 5-ring structures and larger ring systems (VII). An example chromatogram of Synthoil, a hydrodesulfurization process coal liquid, is shown in Figure 1. Saturated compounds eluted earlier than Fraction I and were discarded. Partially saturated compounds (which may constitute substantial constituents of synthetic oils produced by hydrogenation procedures) eluted earlier than their fully aromatic counterparts, usually in the fraction corresponding to the number of aromatic rings; for example, 9,10-dihydrophenanthrene appeared in Fraction II (two aromatic rings) rather than Fraction III. Compounds possessing the partially saturated cyclopentadiene ring (i.e., fluorene derivatives) and polyphenyls also eluted earlier than condensed hydrocarbons having the same number of rings. These compounds may well be split between fractions in the present scheme and, if present in large quantities, probably would require an adjustment in fraction volumes for optimum collection. The effect of ring alkylation

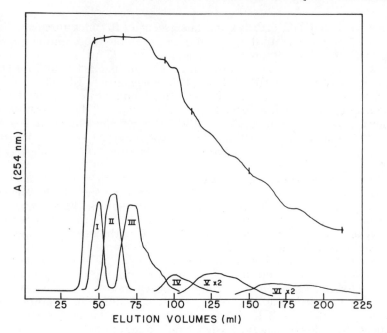

Figure 1.   Excitation wavelengths: 292.5 nm for BaP; 313 nm for BeP.   Upper curve: liquid chromatogram of a sample of Synthoil in isopropyl alcohol; the ticks on the chromatogram denote the various Fractions I through VI.   Lower curves: chromatograms of each of the individual Fractions I through VI after being rechromatographed on the same column.

on elution volume has not been established conclusively but is probably relatively small.

Synthetic fuels may also contain significant concentrations of heterocyclics, amines, and phenols. It appears that nitrogen heteroatoms in ring systems tend to decrease the retention volume of a compound (by comparison with the parent PAH); oxygen heteroatoms appear to have little effect on retention volume, and sulfur heterocyclics tend to exhibit greater retention than the parent hydrocarbons. For example, acridine elutes in Fraction I, whereas anthracene (the parent hydrocarbon) elutes in Fraction III; dibenzofuran elutes at about the same volume as fluorene, while dibenzothiophene elutes later than fluorene. Phenols and amines having protons available for hydrogen bonding to PVP elute much later than the parent PAH. For example, most derivatives of phenol and aniline elute in Fractions III and IV; carbazole elutes in Fraction VI. Much larger amines and phenols may not be sampled by the present LC procedure because of their very long retention times.

General characteristics of the MI FTIR and fluorescence spectra of pure PAH have been described in detail previously (*16, 23, 24*); here we wish only to emphasize the power of MI spectrometry to distinguish between isomeric PAH. Figures 2 and 3, respectively, show the MI FTIR and fluorescence spectra of benzo[a]pyrene and benzo[e]pyrene; in both cases the spectra are quite different, and it would be a straightforward matter to identify either compound in the presence of its isomer. Demonstrations of the ability of MI IR and fluorescence spectrometry to distinguish individual compounds in mixtures of the various four-ring PAH have appeared previously (*16, 23*).

Figures 4, 5, and 6 are MI fluorescence spectra of Fractions IV, V, and VI, respectively, from the LC of a sample of Synthoil, a catalytic hydrodesulfurization process coal liquid. By comparison with MI fluorescence spectra of standard PAH, the following compounds can be identified conclusively in the indicated fractions: pyrene (IV), benz[a]-

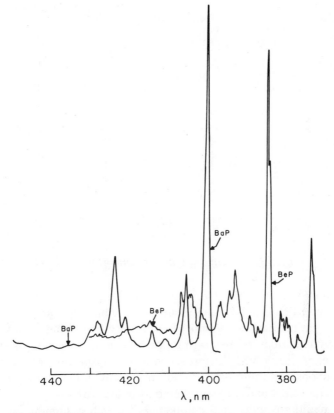

Figure 2. MI (in N₂) fluorescence spectra of 100 ng each of BaP and BeP. These compounds absorb at different wavelengths; hence, different excitation wavelengths must be used to observe their fluorescence spectra.

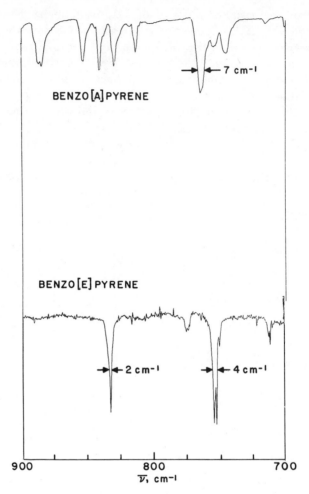

Figure 3. MI (in N_2) FTIR spectra of BaP and BeP

anthracene (V), chrysene (V), and benzo[a]pyrene (VI). Pyrene is found only in Fraction IV; the pyrene bands which appear in the fluorescence spectra of Fractions V and VI are artifacts resulting from vaporization of pyrene from rubber o-rings present in the vacuum system. A number of well-resolved fluorescences of other species are observed in these fractions, but the spectra do not match any of our standard spectra; hence, these compounds are designated as unknown (U).

To quantitate individual compounds in such a sample, an internal standard must be added to the sample (to compensate for the possibility that deposition of the sample on the window may not be quantitative (16)), and standard addition must be performed (to compensate for any absorption of fluorescence of the analyte of interest by other sample constituents). In order to use standard addition, the observed fluorescence intensity must be linear in the quantity of analyte. Figure 7 shows an example working curve for a pure hydrocarbon (chrysene). The curve is linear from the detection limit (20 picograms) to 5 μg; thus, the linear working range is in excess of five decades in chrysene concentration, indicating that standard addition should be capable of yielding reliable quantitative results for PAH. The PAH identified in the various LC fractions of a sample of Synthoil were quantitated by successive standard

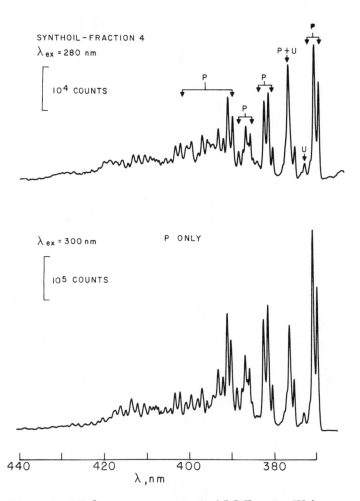

Figure 4.   MI fluorescence spectrum of LC Fraction IV from Synthoil at two excitation wavelengths. P = pyrene; U = unknown.

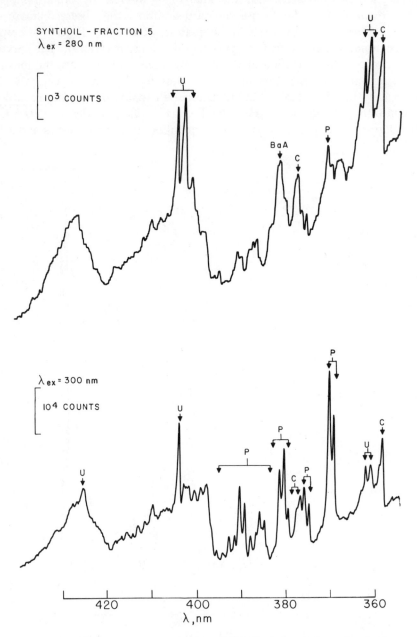

Figure 5.   MI fluorescence spectrum of LC Fraction V from Synthoil at two excitation wavelengths. BaA = benz[a]anthracene; C = chrysene; U = unknown; P = pyrene (artifact; not present in fraction).

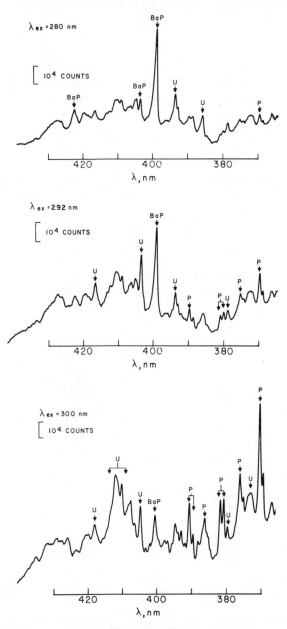

Figure 6. MI fluorescence spectrum of LC Fraction VI from Synthoil at three excitation wavelengths. BaP = benzo[a]pyrene; U = unknown; P = pyrene (artifact; not present in fraction).

addition, using benzo[b]fluorene as internal standard; the results are listed in Table I.

The examination of MI fluorescence spectra of the analogous fractions from the LC of a shale-derived oil (from a simulated in situ burn from Mahogany zone of Green River shale) reveals the presence of the same compounds, in the same fractions, as in the Synthoil fractions. The fluorescence spectra of the shale oil and Synthoil fractions have similar overall appearances. Quantitative analyses for pyrene, benz[a]anthracene (BaA), chrysene, and benzo[a]pyrene (BaP) were performed in the shale oil fractions; the results are listed in Table I.

In view of the well-resolved fine structure in the MI fluorescence spectra of these complex LC fractions, it is evident that the potential information content of these spectra is very great. However, the necessity of matching the observed spectrum of an unknown sample constituent

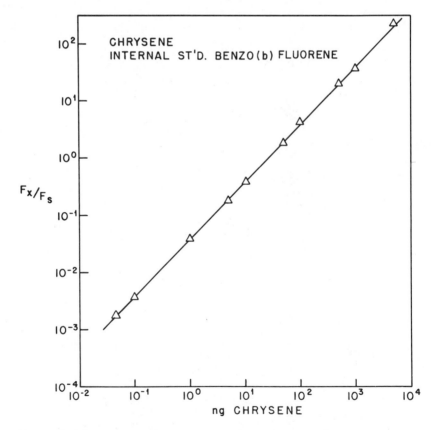

Figure 7.   Quantitative fluorescence working curve for chrysene matrix isolated nitrogen, using benzo[b]fluorene as internal standard

Table I.   Quantitative Analyses for PAH in Oil Samples

Compound	Oil Type	Fraction	Concen-tration[a]	Method[b]
Pyrene	Synthoil	IV	67	IR
			31	F
	shale oil	IV	9.3	IR
			5.6	F
Chrysene	Synthoil	V	0.53	IR
			0.56	F
	shale oil	V	2.0	IR
			2.1	F
Benz[a]anthracene	Synthoil	V	0.67	F
	shale oil	V	0.93	F
Benzo[a]pyrene	Synthoil	VI	0.73	F
	shale oil	VI	1.67	F

[a] As ppm in original oil sample.
[b] IR = MI FTIR spectrometry; F = MI fluorescence spectrometry.

with a standard spectrum of a pure compound to achieve a positive identification is a severe shortcoming of fluorescence spectrometry. (Based upon changes in the appearance of the spectra with changing excitation wavelength, it appears that there are, in this particular Synthoil sample, at least two unidentified fluorescent compounds in Fraction IV, three unidentified fluorescent compounds in Fraction V, and six unidentified fluorescent compounds in Fraction VI.) The unavailability of suitable reference samples of many PAH poses a formidable obstacle to identification of these compounds by fluorometry.

Fractions from the LC of Synthoil and shale oil were examined also by MI FTIR spectrometry; spectra of Fractions IV and V from the LC of Synthoil are shown in Figures 8 and 9, respectively. Pyrene ($v = 750, 746$ cm^{-1}) and chrysene ($v = 766, 764, 817, 814$ cm^{-1}) can be identified as present in Fractions IV and V, respectively. The other absorption bands in the FTIR spectra of Fractions IV and V cannot be assigned to the PAH for which MI spectra are available. No absorptions assignable to PAH are detected in the MI FTIR spectra of Fraction VI of Synthoil. The presence of strong absorption bands in the 2920–3000 cm^{-1} in the IR spectra of the oil fractions is indicative of the presence of alkyl groups, and it is therefore likely that the high concentrations of paraffin hydrocarbons largely obscure the IR features of PAH (presumably present in much smaller quantities). Thus, a pretreatment step (25) prior to the LC of the oil samples, to remove aliphatics, appears essential in the examination of these samples of PAH by FTIR spectrometry.

The IR spectra of all LC fractions from both Synthoil and shale oil contain a very intense feature at 750 cm⁻¹ and a less intense band at 820 cm⁻¹. These bands, together with a strong feature at 3495 cm⁻¹, are indicative of N–H stretching vibrations, thus implying the presence of amines and/or N-heterocyclics in substantial quantities in these samples. Indeed, previous analyses (26) of Synthoil have demonstrated the presence of relatively high concentrations of indole (210 ppm) and 3-methylindole (130 ppm). However, a careful examination of MI fluorescence spectra of indole and a number of substituted indoles shows no match with any of the major unknown bands in the MI fluorescence spectra of the various Synthoil and shale oil fractions (27). Thus, it remains to identify the unknown fluorescent constituents of these samples, and the identities of the compounds responsible for the intense N–H absorptions in the FTIR spectra remain to be established.

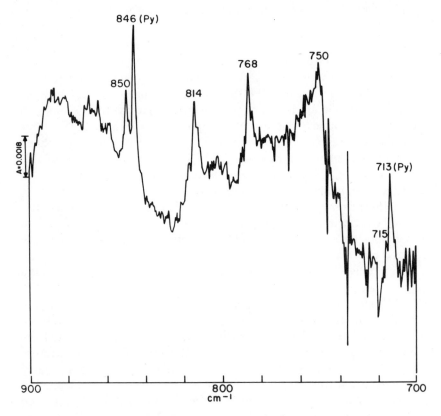

Figure 8. FTIR absorption spectrum of Synthoil LC Fraction IV in a nitrogen matrix. P = pyrene; U = unknown.

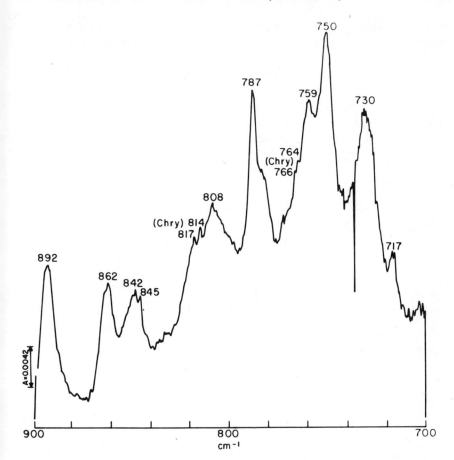

Figure 9.   FTIR absorption spectrum of Synthoil LC Fraction V in a nitrogen matrix. Chr = chrysene.

Performance of quantitative analyses by IR spectroscopy is dependent upon adherence, by the absorbing compound, to Beer's law over a significant concentration range. Indeed, one of the historical impediments to use of IR for quantitative analysis of PAH in real samples was the failure to observe linear Beer's law behavior in liquid solutions or KBr discs at the high concentrations required to achieve detectable signals using conventional dispersive IR spectrometers (21). By using MI sampling and FTIR instrumentation, the range of applicability of Beer's law can be expanded greatly. Figure 10 shows an example of a Beer's-law plot for a pure PAH (perylene) in a nitrogen matrix. By the combined use of two different MIIR sampling systems (including an ultramicro sampling system (28) which extends the FTIR detection limit for perylene to 50 ng, or 2×10^{-10} mol), linear Beer's-law plots can be achieved over more than three decades in PAH concentration. Thus, successive standard addition can be used for quantitation of individual compounds in mixtures by MI IR spectrometry.

The results of quantitative analyses for pyrene and chrysene in the Synthoil and shale oil samples are compared with those obtained for the same samples by MI fluorescence spectrometry in Table I. Several relevant conclusions emerge. First, both oils contain significant concentrations of the highly carcinogenic compound benzo[a]pyrene; the shale oil sample contains nearly 2 ppm of this hazardous substance. Second,

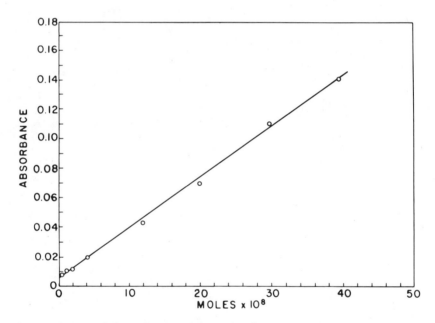

Figure 10. Beer's-law plot for MI FTIR analysis of perylene (resolution = 2.0 cm⁻¹). The 818-cm⁻¹ absorption band of perylene was used for all absorbance measurements.

both oils are rich in pyrene; Synthoil seems to contain about ten times as much pyrene, on a weight basis, as does the particular shale oil sample examined in this study. Third, for the compounds capable of detection by both IR and fluorescence, the quantitative results for the two methods were in excellent agreement for chrysene but differed by about a factor of two for pyrene. Inasmuch as pyrene was present in high concentration, especially in Synthoil, it is likely that the results for fluorescence analysis were perturbed by reabsorption effects (*12*); hence, it is likely that the FTIR results for pyrene are more accurate. (In samples of this complexity, it is obviously highly desirable to use two or more independent quantitative measurements, in order to detect large systematic errors.)

Conclusion

We have demonstrated that the relatively simple LC system, using PVP as stationary phase, provides useful fractionation for complex mixtures of PAH, and that the LC fractions so obtained can be analyzed, both qualitatively and quantitatively, for individual PAH by matrix isolation spectroscopy. These procedures are quite new, and obviously additional work will be required before they can be used in the routine characterization of coal liquids and shale oils. The LC procedure must be modified in order to eliminate the aliphatics, which interfere severely in the IR analysis; the amines and/or nitrogen hetero aromatics which appear to be present in these samples must be identified; the unknown species present in the fluorescence spectra of the various fractions must be identified; and the inner-filter effects which appear to have produced low fluorometric results for pyrene (Table I) must be eliminated (by using higher matrix-to-sample ratios than those used customarily (16) in MI fluorometry). In view of the complexity of the samples, and the differing nature of the analytical information provided by the various spectroscopic techniques (electron impact and chemical ionization mass spectrometry, fluorescence and phosphorescence, IR, NMR, etc.), it is apparent that a sophisticated analytical plan for full characterization of synthetic fuels will necessarily utilize the full range of fractionation and analytical methods available. The methods described in this paper are, in large measure, complementary to (rather than competitive with) more established procedures, such as GC–MS. The combination of LC and MI spectroscopy should in due course assume a role as one of the several combined separation–high resolution spectrometric techniques useful for the identification and quantitation of individual compounds present at low levels in very complex samples, such as coal liquids and shale oils.

Acknowledgment

This research is supported in part by the Electric Power Research Institute (Contract 741011–RP–332–1) and the National Science Foundation (Grant MPS75–05364). The purchase of the FTIR spectrometer was supported in part by a National Science Foundation Research Instrument Grant (GP–41711).

Oak Ridge National Laboratory is operated by the Union Carbide Corporation for the U.S. Energy Research and Development Administration.

Literature Cited

1. Lao, R. C., Thomas, R. S., Oja, H., Dubois, L., *Anal. Chem.* (1973) **45,** 908.
2. Janini, G. M., Muschik, G. M., Schroer, J. A., Zielinski, W. L., Jr., *Anal. Chem.* (1976) **48,** 1879.
3. Lee, M. L., Hites, R. A., *Anal. Chem.* (1976) **48,** 1890.
4. Scheppele, S. E., Grizzle, P. L., Greenwood, G. J., Marriott, T. D., Perreira, N. B., *Anal. Chem.* (1976) **48,** 2105.
5. Aczel, T., Williams, R. B., Pancirov, R. J., paper presented at Pittsburgh Conference on Analytical Chemistry and Applied Spectroscopy, Cleveland, Ohio, February 28, 1977.
6. Hase, A., Lin, P. H., Hites, R., "Polynuclear Aromatic Hydrocarbons: Chemistry, Metabolism, and Carcinogenesis," R. Freudenthal and P. W. Jones, Eds., p. 435, Raven, New York, 1976.
7. Leo, R. C., paper presented at Second Annual Workshop on Polynuclear Aromatic Hydrocarbons, Oak Ridge, Tennessee, March 11, 1977.
8. Cautreels, W., Van Cauwenberghe, K., *Atmos. Environ.* (1976) **10,** 447.
9. Goldstein, G., *J. Chromatogr.* (1976) **129,** 61.
10. Sawicki, E., *Talanta* (1969) **16,** 1231.
11. Richardson, J. H., Wallin, B. W., Johnson, D. C., Hrubesh, L. W., *Anal. Chim. Acta* (1976) **86,** 263.
12. Parker, C. A., "Photoluminescence of Solutions," pp. 72–77, 220–222, 344–350, 440–443.
13. Shpol'skii, E. V., Bolotnikova, T. N., *Pure Appl. Chem.* (1974) **37,** 183.
14. Kirkbright, G. F., de Lima, C. G., *Analyst (London)* (1974) **99,** 338.
15. Lukasiewicz, R. J., Winefordner, J. D., *Talanta* (1972) **19,** 381.
16. Stroupe, R. C., Tokousbalides, P., Dickinson, R. B., Jr., Wehry, E. L., Mammantov, G., *Anal. Chem.* (1977) **49,** 701.
17. Wehry, E. L., *Fluoresc. News* (1974) **8,** 21.
18. Meyer, B., "Low Temperature Spectroscopy," American Elsevier, New York, 1971.
19. Cradock, S., Hinchcliffe, A. J., "Matrix Isolation: A Technique for the Study of Reactive Inorganic Species," Cambridge University, New York, 1975.
20. Moskovits, M., Ozin, G. A., "Cryochemistry," Wiley, New York, 1976.
21. "Particulate Polycyclic Organic Matter," p. 197–298, National Academy of Sciences, Washington, D. C., 1972.
22. Griffiths, P. R., "Chemical Infrared Fourier Transform Spectroscopy," Wiley, New York, 1975.
23. Mamantov, G., Wehry, E. L., Kemmerer, R. R., Hinton, E. R., *Anal. Chem.* (1977) **49,** 86.
24. Wehry, E. L., Mamantov, G., Kemmerer, R. R., Brotherton, H. O., Stroupe, R. C., "Polynuclear Aromatic Hydrocarbons: Chemistry, Metabolism, and Carcinogenesis," R. Freudenthal and P. W. Jones, Eds., p. 299, Raven, New York, 1976.
25. Novotny, M., Lee, M. L., Bartle, K. D., *J. Chromatogr. Sci.* (1974) **12,** 606.
26. Shults, W. D., Ed., "Preliminary Results: Chemical and Biological Examination of Coal-Derived Materials," Oak Ridge National Laboratory, 1976.
27. Tokousbalides, P., unpublished results, University of Tennessee (1977).
28. Kemmerer, R. R., Ph.D. thesis, University of Tennessee, Knoxville (1977).

RECEIVED August 5, 1977.

Chromatographic Studies on Oil Sand Bitumens

M. SELUCKY, T. RUO, Y. CHU, and O. P. STRAUSZ

Hydrocarbon Research Center, Department of Chemistry, University of Alberta, Edmonton, Alberta, Canada T6G 2G2

A rapid chromatographic procedure is described using silica gel for the separation of deasphaltened bitumen. Comparison of the results of silica gel separations with those from the USBM API–60 procedure shows that contamination of the hydrocarbon fraction by polar materials caused by the omission of resin-retaining columns is slight and the method may be satisfactory for a number of applications. The ability of silica gel to separate polar material from hydrocarbons has been exploited for a rapid gravimetric method for the determination of polar materials in bitumens. Finally, a gas chromatographic/high pressure liquid chromatographic (GC/HPLC) fingerprinting procedure is outlined, based on silica gel separation of hydrocarbon classes for the fast characterization of petroleum-derived materials.

The last few decades have witnessed a major effort on the part of analytical chemists to develop suitable analytical methods for the analysis of crude oils, culminating in the development of a detailed analytical procedure utilizing a broad variety of modern instrumental techniques known to the petroleum chemist as the USBM–API 60 method (1). However, the utility of this method is constrained to some extent because it is rather time consuming and generally is not suited for the analysis of large numbers of samples.

In the course of our studies on Athabasca and Cold Lake bitumens, the problem of determining the overall composition of a number of samples from cores and from various extraction procedures made it necessary to sacrifice some analytical detail in favor of speed of analysis. Since no comprehensive analytical data on either bitumen were available, we made a comparative study of a separation procedure based on the USBM–API 60 separations and on a simple silica/alumina separation of deasphaltened bitumens. The results of these studies have been reported elsewhere (2, 3). These comparative studies were done also with the aim of developing a simple and satisfactory separation scheme amenable to conversion in the HPLC mode which would allow treatment of large numbers of samples for geological and geochemical studies. The extent to which the omission of the ion exchange steps influence the quality of separation also had to be established.

In view of the high contents of total polar and high molecular weight material representing roughly 50% of these bitumens, an attempt was made to develop a simple, inexpensive, and fast method for the direct determination of the bulk of this material prior to the HPLC separation, based on the straightforward separation of the deasphaltened bitumen on silica. The removal of the total polar material simplifies the calibration of HPLC chromatograms (4, 5, 6).

In the course of the work, it was noted that HPLC chromatograms can be used also for fingerprinting purposes. Combination of this method with GC was applied successfully to the combined fingerprinting of unknowns such as spills, contents of tanks, fast checking of gasolines in filling stations, etc. Details of this new fingerprinting technique and meaningful interpretation thereof will be reported elsewhere (7).

Results and Discussion

Table I summarizes the analytical results for deasphaltened Athabasca bitumen (without prior distillation) on a series of columns used in the USBM-API 60 procedure, and those obtained by the simplified silica and alumina class separation scheme. As seen, the results are comparable provided that the polyaromatic and polar fractions are combined. The total analyses of the separated fractions obtained by the two methods listed in Table II are also in good agreement, with the exception that sulfur values from the simplified procedure are somewhat higher in all aromatic fractions except the polyaromatic/polar fraction.

Since in the simplified procedure only the hydrocarbon fraction from silica is reseparated on alumina, while in the extended procedure all the material remaining after removal of the saturates is rechromatographed on alumina, the contents of polyaromatics from the former is necessarily lower (9.4%) than from the latter (28.6%) procedure. The balance of material is represented in the simplified procedure by fraction Polar 1 (see also References 2, 3, and below).

On the other hand, analysis for sulfur and oxygen shows that in the polyaromatic/polar fractions, of an av mol wt of 490, 6.8% sulfur and 2.16% oxygen represent on the average 1.04 sulfur atoms and nearly 0.7 oxygen atoms per molecule. The material is therefore essentially heterocyclic in character. The nitrogen contents of the fractions were also comparable, suggesting that in the class separation method most of the

Table I. Class Separations of Athabasca Bitumen

Separated Fraction	Procedure	
	n-C5	n-C5
	ion exchangers Fe^{+++}/clay SiO_2 Al_2O_3	SiO_2 Al_2O_3
	Wt %	Wt %
Asphaltenes	16.6	16.6
Saturates	20.2	21.4
Monoaromatics	9.9	8.8
Diaromatics	4.2	4.5
Polyarom. + nonspec. polar	26.3 ⎫	
acids	14.6 ⎬ 49.2	48.8
bases	6.7 ⎭	
neutral nitrogen compounds	1.6	

nitrogen compounds are retained on silica, whereas in the extended procedure, most of them were retained on the ferric chloride column.

The separation scheme for deasphaltened bitumen on silica [WOELM, activated at 140°C (4 hr)] is shown in Figure 1. The fractions obtained were loosely termed hydrocarbons, Polar I, II, and III. The IR spectra of these fractions, as shown in Figure 2, suggested that Polar I was very similar to the polyaromatic/neutral polar fraction from the API 60-based separation after removal of additional material from alumina with pure benzene. Also, simulated distillation curves for this fraction from either procedure are very similar (Figure 3). The IR spectra of the Polar II and III fractions show the presence of all the functional groups which can be distinguished in these complex mixtures and which

Table II. Total Analysis of Hydrocarbon Fractions of Athabasca Oil

Fraction	%					Molecular Weight	H/C
	Carbon	Hydrogen	Nitrogen	Oxygen	Sulfur		
Saturates	85.02[a]	12.92	0.0	0.0	0.17	365	1.82
	86.67[b]	13.17	0.0	0.0	0.15	365	1.82
Monoaromatics	87.71[a]	11.69	0.0	0.60	0.26	340	1.60
	86.85[b]	11.01	0.0	0.41	1.71	345	1.53
Diaromatics	87.62[a]	10.56	0.0	0.57	2.39	380	1.45
	86.13[b]	10.42	0.0	0.50	2.94	420	1.45
Polyaromatics	81.31[a]	9.79	0.13	1.98	7.07	490	1.44
	81.34[b]	9.68	0.0	2.16	6.81	505	1.43

[a] Extended procedure.
[b] Straightforward procedure.

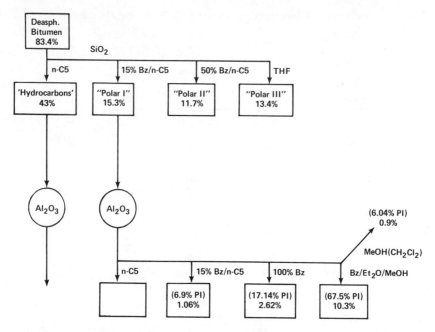

Figure 1. Separation of deasphaltened bitumen on silica and of Polar I fraction on alumina. The numbers in brackets are % of sample, those outside brackets are % bitumen.

have been discussed in some detail by Bunger (8). The total percentage of the Polar II and III fractions (25.1%) was slightly higher than that of combined acids, bases, and neutral nitrogen compounds from the API 60-based separation (22.9%), while the total polyaromatic/polar material was isolated in approximately equal amounts from either procedure (48.8% and 49.2% respectively, cf. Table I). The analyses of the polar fractions obtained from the straightforward silica separation are presented in Table III together with the results for the polyaromatic/neutral polar fraction from the API 60 procedure. Comparative data for other fractions obtained from the two procedures are presented in Table II.

The Polar 1 fraction was separated further on alumina (zero water, *see* Figure 1). Elution with 15% Bz/nC_5 (which elutes monoaromatics

Table III.   Fractions Eluted from Silica[a]

Fraction	Car-bon	Hydro-gen	Oxy-gen	Nitro-gen	Sul-fur	Molecu-lar Weight	H/C
Hydrocarbons	85.93	12.30	0.21	0.10	2.14	325	1.72
Polar I	83.87	9.75	0.60	0.43	5.96	485	1.40
Polar II	82.19	9.53	1.63	0.81	5.61	840	1.39
Polar III	76.34	9.35	7.21	0.96	5.81	740	1.47

[a] Cold Lake, cold-bailed bitumen.

and the bulk of diaromatics) gave an additional 1.06% material, while 100% benzene (which elutes tri- and higher aromatics) gave an additional 2.62% material together with 10.3% of material elutable with $Bz/Et_2O/MeOH(60/20/20)$. These results show clearly that omission of ion exchangers and the Fe^{+++}/clay column does not substantially impair the analysis for hydrocarbons.

It was also of interest to establish the fate of acids, bases, and neutral nitrogen compounds on silica and alumina. For this purpose, the acids, bases, and neutral nitrogen compounds isolated on ion exchangers and Fe^{+++}/clay, respectively, were applied on a silica column. Elution of hydrocarbons with *n*-pentane yielded 0.29% material, which was separated on alumina into mono + diaromatics (0.1%) and polyaromatics (0.24%) (Figure 4). Thus, if ion exchangers and the Fe^{+++}/clay column are omitted, there remains only about 0.8% impurities stemming from acid and base fractions present in the total mono- and diaromatic fractions.

Elution from the silica gel column with tetrahydrofuran (THF) gave 22.14% material, 0.43 being retained. All of the above data are expressed as percent of bitumen. The total recovery was 98.11% of total acids + bases + nitrogen compounds which, again, is satisfactory considering that the recoveries from the API 60-based separations range between 95 and 97%. Thus, there is very little irreversible adsorption of high molecular weight material. In fact, the recoveries could be improved by deactivating

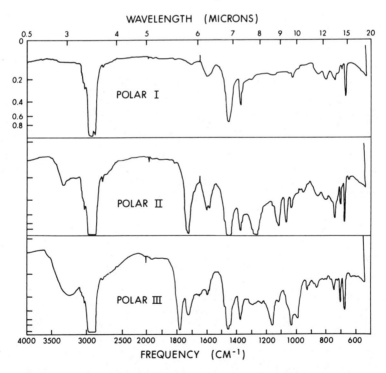

Figure 2.   IR spectra of Polar fractions from the silica separation

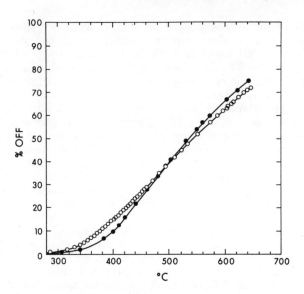

Figure 3. Simulated distillation of polyaromatic/ polar fraction from API-60 (○) and of Polar I fraction (●)

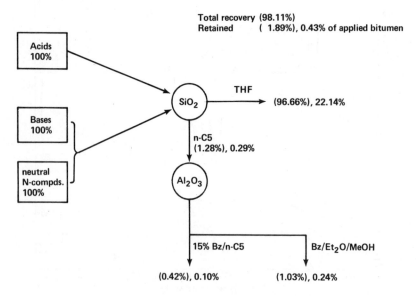

Figure 4. Separation of acids and bases from ion exchange on silica and alumina

the silica column with a stronger solvent (e.g. methanol) and eluting the methanol-insoluble material with a solvent proper, e.g. methylene chloride.

The separations discussed above have shown clearly that omission of the ion-exchange and complexation steps bring about very little impairment of the quality of hydrocarbon separations and can be used safely with heavy bitumens (as, e.g. the bitumens from the Athabasca region) without prior distillation in order to speed up the course of analysis

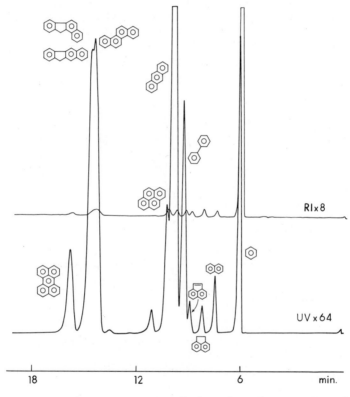

Figure 5. Elution of aromatic hydrocarbons from a μ-Porasil column. 60 cm × 1/4"; n-heptane eluent; P = 1200 psi; F = 1.5 mL/min.

unless, of course, specific information about the contents of acids, bases, and neutral nitrogen compounds is being sought.

Furthermore, these results, together with the ease of elution of many tetra- and pentacyclic polyaromatic hydrocarbons from μ-Porasil columns, as illustrated in Figure 5, confirm the potential of the published HPLC method for petroleum-derived products (4, 5, 6) where large numbers of bitumen samples are involved.

Whenever this type of analysis is satisfactory, the HPLC method can be simplified further by removing the bulk of hydrocarbons from the essentially heterocyclic material prior to HPLC analysis. The insertion of a precolumn to withdraw resins has been described (9). In this chapter the possibility of gravimetric determination of total polar material on a short column of silica is reported. Short pieces of glass tubing were packed with known amounts of silica gel activated at 140°C (200 mm Hg for 1 hr) and weighed. The sample was applied and the tube reweighed. After elution with anhydrous n-pentane which, if pressure assisted, takes about 30 min, the residual pentane was removed from the tube packing with a stream of dry nitrogen and the tube was reconditioned under exactly the same conditions under which it had been preconditioned. The reconditioned tube was weighed and the weight loss corresponds to the weight of eluted hydrocarbons, while the weight difference between the sample-containing tube after elution and the tube itself corresponds to

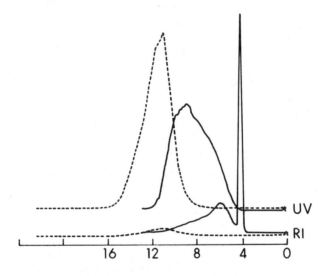

Figure 6. High-performance liquid chromatogram of hydrocarbon fractions eluted from tubes: first 45-mL eluate (——); second 50-mL eluate (– – –)

the weight of adsorbed polar material. The difference between the sample weight and the combined weights of hydrocarbons and polar material corresponds to the water content of the sample.

The analysis worked well for samples containing up to 10% water. At higher water contents, the repeatability deteriorated with increasing water content. The reproducibility of the results, e.g., constancy of tube weight upon removal of solvent and volume of eluent needed for complete hydrocarbon elution were checked, the latter by HPLC on μ-Porasil. Figure 6 shows a typical course of the hydrocarbon elution. Comparison with data obtained from preparative separations (Table IV) showed good agreement for the bitumen samples. The applied sample size is limited by the weight of silica used to pack the tubes. Using ∼ 4 g of silica, a breakthrough of colored material occurred when the sample exceeded 50

Table IV. **Gravimetric Determination of Total Polar Material in Athabasca Bitumen**[a]

Sample	% Polar	
	Incomplete Elution (45 mL eluent)	*Complete Elution (80 mL eluent)*
1	59.5	54.5
2	57.2	56.7
3	59.5	56.3
	Av 58.7 ± 1.0	55.8 ± 0.9

[a] The preparative procedure yielded av 56.35% (deasphaltened sample).

mg. This amount of sample is, however, more than satisfactory for subsequent HPLC separation of hydrocarbons and high-voltage, low-resolution MS, GC, NMR, and UV spectroscopy.

At the same time, this procedure simplifies quantification of HPLC chromatograms of the hydrocarbon fraction since two unknowns—polar material and water—have been eliminated. Furthermore, the total amount of hydrocarbons can be determined easily by gravimetry. This

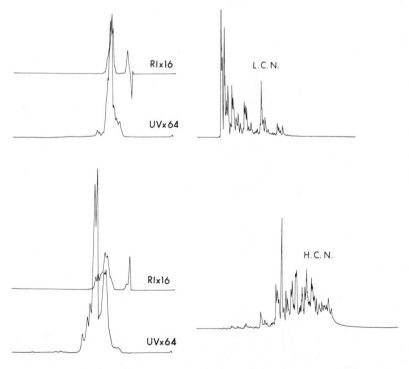

Figure 7.   Combined fingerprints of light-(LCN) and heavy-(HCN) cracked naphthas

method is simple, relatively fast (it can be speeded up by pumping the eluent if the tube assembly is slightly modified), and inexpensive, and allows concurrent analyses of large numbers of samples.

HPLC on a high-efficiency silica column can be used also for finger-printing unknown materials. Fingerprinting techniques have been used in many fields of applied analysis and in the present work we have shown that combined GC/HPLC (silica) fingerprinting using RI and UV detectors in tandem will furnish considerable information on boiling point distribution, compound class distribution, and other characteristics of an unknown sample. The technique can be used for various petroleum fractions, examples of which are shown in Figures 7 and 8, and even for the comparison of crudes. The HPLC traces illustrated in Figure 9 clearly show that the saturate and monoaromatic fractions become rapidly depleted with increasing biodegradation. Thus while GC by itself only gives information on the boiling range of the sample, additional informa-tion on its composition in terms of hydrocarbon classes thus enhances the potential of fingerprinting. Details of meaningful interpretation of combined fingerprinting are reported elsewhere (7). We have success-

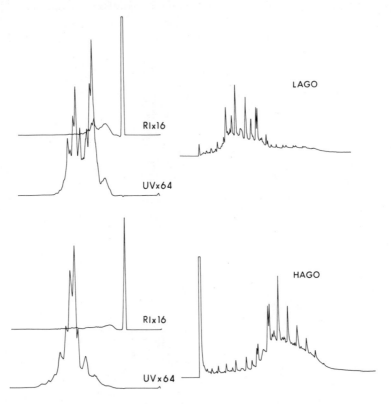

Figure 8.   Combined fingerprints of light-(LAGO) and heavy-(HAGO) atmospheric gas oils

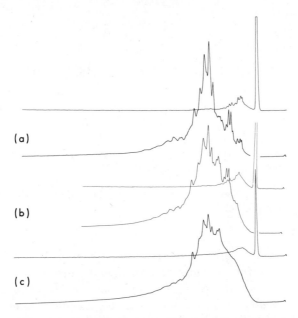

Figure 9.   HPLC fingerprints of nonbiodegraded (a), partially degraded (b) and biodegraded (c) bitumens

fully used this combined method to characterize a number of samples from in situ and other cracking procedures and were able to correctly deduce their history in ~ 1 hr.

Acknowledgments

The authors thank the National Research Council of Canada for continuing financial support and Dr. E. M. Lown for helpful assistance.

Literature Cited

1. Haines, W. E., Thompson, C. T., "Separating and Characterizing High Boiling Petroleum Distillates: The USBM–API Procedure," LERC/RI–75/15, BERC/RI–75/2, July, 1975.
2. Selucky, M. L., Chu, Y., Ruo, T. C. S., Strausz, O. P., *Fuel* (1977) **56**, 369.
3. Selucky, M. L., Chu, Y., Ruo, T. C. S., Strausz, O. P., *Fuel* (1978) **57**, 9.
4. Suatoni, J. C., Swab, R. E., *J. Chromatogr. Sci.* (1975) **13**, 361.
5. Suatoni, J. C., Garber, H. R., *J. Chromatogr. Sci.* (1976) **14**, 546.
6. Suatoni, J. C., Swab, R. E., *J. Chromatogr. Sci.* (1976) **14**, 535.
7. Selucky, M. L., Ruo, T. C. S., Strausz, O. P., *Fuel*, in press.
8. Bunger, J. W., *Am. Chem. Soc., Div. Pet. Chem., Prepr.* (1977) **22**, 716.
9. Jewell, D. M., Albaugh, E. W., Davis, B. E., Ruberto, R. G., *Ind. Eng. Chem., Fundam.* (1974) **13**(3), 278.

RECEIVED August 5, 1977.

9

Petroleum Asphaltenes: Chemistry and Composition

J. F. McKAY, P. J. AMEND, T. E. COGSWELL, P. M. HARNSBERGER, R. B. ERICKSON, and D. R. LATHAM

Energy Research and Development Administration, Laramie Energy Research Center, Laramie, WY 82071

An n-pentane asphaltene prepared from a Wilmington, California crude oil was separated into fractions of acids, bases, neutral nitrogen compounds, saturate hydrocarbons, and aromatic hydrocarbons using ion exchange, coordination, and adsorption chromatography. Major compound types identified in the acid and base fractions by IR spectrometry include carboxylic acids, phenols, amides, carbazoles, and pyridine benzologs. The average molecular weight of the asphaltene was estimated to be between 500 and 800 by vapor-pressure osmometry (VPO), low-resolution mass spectrometry (MS), and quantitative IR analyses. The results indicate that the asphaltene is a complex mixture of the most polar and highest molecular weight molecules of the crude oil. Asphaltene precipitation is explained as being a solubility phenomenon. When n-pentane is added to the crude oil, the solvent properties and average molecular weight of the system are changed so that the most polar and highest molecular weight molecules are no longer soluble and thus precipitate as asphaltenes.

Deasphaltening of petroleum is the well-known process of treating petroleum with a low molecular weight hydrocarbon solvent to remove the asphaltene component of petroleum that interferes with refining of petroleum. An asphaltene must be defined by the solvent used to precipitate it since different solvents cause different amounts of precipitation. For example, an *n*-pentane asphaltene is that material precipitated from petroleum when a large volume of *n*-pentane is added to the petroleum sample. The amount of asphaltene precipitated depends primarily on the hydrocarbon solvent (1, 2), the volume ratio of solvent to petroleum (2), and the composition of the petroleum (3, 4). Asphaltenes generally are regarded as high molecular weight (1,000–50,000) materials (2, 5) that contain large amounts of nitrogen, sulfur, and oxygen compounds. An understanding of the chemistry and composition of petroleum asphaltenes is important in the total analyses of a petroleum and should be useful for the design of efficient catalysts for the refining of petroleum. In addition, this information will be useful to researchers working with synthetic fossil fuel mixtures such as coal liquids, tar sand bitumens, and shale oil.

Many questions concerning the nature of petroleum asphaltenes remain unresolved: (1) What is the chemical composition of petroleum asphaltenes? (2) What are the molecular weights of asphaltene components? (3) Why are asphaltenes precipitated from solution in petroleum by the addition of a hydrocarbon solvent such as n-pentane? In this chapter we attempt to answer these questions. In addition, we suggest that asphaltene formation is a general phenomenon that is pertinent to the chemistry of coals, tar sand bitumens, shale oil, and other complex solutions of organic compounds.

This paper discusses the initial experiment made to isolate n-pentane asphaltenes from a Wilmington, California crude oil and the separation of the asphaltene into fractions of acids, bases, neutral nitrogen compounds, saturate hydrocarbons, and aromatic hydrocarbons, using ion exchange chromatography, coordination chromatography, and adsorption chromatography. The major compound types in the acid fraction have been identified by IR spectrometry. Molecular weights of compound types were estimated by VPO, mass spectrometry, and quantitative IR spectrometry.

Experimental

Reagents. Amberlite IRA 904 and Amberlyst 15, the anion- and cation-exchange resins, were obtained from Rohm & Haas. Attapulgus clay (LVM, 50/80 mesh) was obtained from Engelhard Minerals and Chemicals Corp., and the silica gel (grade 62, 60/200 mesh) came from Davison Chemical Co. Reagent-grade ferric chloride hexahydrate and potassium hydroxide were obtained from Baker and Adamson Co. Cyclohexane and n-pentane (99%, Phillips Petroleum) were purified by flash distillation and by percolation through activated siliga gel; benzene, methanol, and 1,2-dichloroethane (reagent grade, Baker and Adamson) were flash-distilled; isopropyl amine (reagent grade, Eastman) was used as received.

Apparatus. The separations were made on a liquid chromatographic column 1.4-cm i.d. × 119-cm long. The column was water-jacketed and contained a recycling arrangement that permits the continuous elution of the sample without the need for large quantities of solvent. Solvent was removed from the fractions using a rotary evaporator and a nitrogen gas sweep at steam bath temperature until a constant sample weight was obtained. IR spectra were recorded in methylene chloride solvent, using a Perkin–Elmer Model 621 IR spectrophotometer; low-resolution mass spectra were recorded on a Varian CH-5 single-focusing MS.

Preparation of Petroleum Asphaltenes. Wilmington, California crude oil (100.7 g) was agitated with n-pentane (1500 mL) at room temperature for 10 min, left unagitated for 1 hr, agitated again for 10 min, and then left unagitated for 15 hr. The solution was decanted from the precipitated asphaltenes. The asphaltenes were filtered using Whatman No. 1 filter paper, washed with n-pentane (200 mL) to remove adsorbed compounds, dried, and weighed. The dried asphaltenes (8.8 g) represented 8.7% of the total crude oil.

Preparation of Resins and Adsorbents. ANION-EXCHANGE RESIN. Amberlite IRA 904 resin (1000 g) was placed in a glass column and activated by the following procedure. The initial washes were made with 1N hydrochloric acid (7.8 L) and distilled water (2.0 L), using a flow rate of 8 bed volumes per hr. The resin was activated using 1N sodium hydroxide (7.8 L). This washing sequence was repeated, starting with hydrochloric acid. The resin then was washed with distilled water (2.0 L). Final preparation of the resin was made by washing with the following solvent sequence: 75% water–25% methanol (1.0 L); 50% water–50% methanol (1.0 L); 25% water–75% methanol (1.0 L); methonal (2.0 L); benzene (3.0 L); cyclohexane (3.0 L). The resin was stored under cyclohexane .(It is important that the resin is not allowed to dry or to be exposed to heat.)

CATION-EXCHANGE RESIN. Amberlite 15 resin was prepared in the manner described for the anion resin except that the acid–base washing sequences were reversed.

FERRIC CHLORIDE ON ATTAPULGUS CLAY. Ferric chloride hexahydrate (40 g), a 10% solution in methanol, was contacted with Attapulgus clay (252 g) for 8 hr. The ferric chloride–Attapulgus clay was filtered, washed several times with cyclohexane, extracted with cyclohexane for 24 hr in a Soxhlet extractor to remove nonadsorbed metallic salt, and dried at room temperature. The material contained 0.7–2.0 wt % of iron.

SILICA GEL ADSORBENT. The silica gel was used as received.

Separation Procedure. The petroleum asphaltenes were separated into five fractions: acids, bases, neutral nitrogen compounds, saturate hydrocarbons, and aromatic hydrocarbons. Acids were isolated using anion-exchange resin, bases with cation-exchange resin, and neutral nitrogen compounds by complexation with ferric chloride adsorbed on Attapulgus clay. The remaining hydrocarbon fraction is separated on silica gel to produce saturate and aromatic hydrocarbon fractions.

ANION-EXCHANGE CHROMATOGRAPHY. A sample of asphaltene (2.7 g) was dissolved in benzene (25.0 mL), and the solution was diluted to 1,000 mL with cyclohexane. No precipitation was observed. The solution was charged to the anion resin (60 g) that had been wet-packed in the column. A small amount of precipitate, estimated to be $\sim 1\%$ of the sample, was observed at the top of the column. Unreactive material was washed from the resin with cyclohexane (200 mL) for ~ 12 hr, using the recycling arrangement of the column. After the unreactive materials were removed, the reactive compounds (acids) were recovered in three subfractions. The first was eluted with benzene. The second subfraction was eluted with 60% benzene–40% methanol (azeotrope). The resin then was removed from the column and placed in a batch apparatus fashioned from a Soxhlet extractor. The third subfraction was recovered by several elutions of 200 mL of 60% benzene–40% methanol saturated with carbon dioxide, each followed by elution with 200 mL of benzene. Upon completion of the separation experiment the small amount of precipitate remained at the top of the column. The three elution steps remove compounds of increasing acid strength.

CATION-EXCHANGE CHROMATOGRAPHY. The sample of acid-free asphaltene, dissolved in 98% cyclohexane–2% benzene, was charged to the A-15 resin (60 g) that had been wet-packed in the column. Unreactive material was washed from the resin with cyclohexane for ~ 12 hr. The reactive material (bases) was recovered from the resin in three subfractions. The first subfraction was removed with benzene. The second subfraction was removed with 60% benzene–40% methanol. The third subfraction was removed in a batch apparatus, using 54% benzene–36% methanol–10% isopropyl amine.

FERRIC CHLORIDE COORDINATION CHROMATOGRAPHY. Ferric chloride–Attapulgus clay (30 g) suspended in cyclohexane was wet-packed in a column. A sample of acid- and base-free asphaltene (0.163 g), dissolved in cyclohexane, was percolated slowly through the column. The entrained oil was removed by 24-hr elution with cyclohexane. The first subfraction of neutral nitrogen compounds was desorbed from the clay by 60- to 72-hr elution with 1,2-dichloroethane. The nitrogen compound–ferric chloride complexes in this fraction were broken by passing the 1,2-dichloroethane solution over anion-exchange resin contained in a second column. The ferric chloride salt was retained on the resin and the nitrogen compounds were recovered in the eluate. A second subfraction was removed from the clay by 60- to 72hr elution with 45% benzene–5% water–50% ethanol. The solvent was removed, and the organic compounds were redissolved in benzene and filtered to remove inorganics; the solvent was removed to eliminate traces of water and ethanol. The organics were redissolved in benzene and passed over anion-exchange resin to remove ferric chloride. The nitrogen compounds were recovered in the benzene eluate. The two subfractions were combined to give a total neutral nitrogen fraction.

SILICA GEL CHROMATOTRAPHY. The acid-, base-, and neutral-nitrogen-free asphaltene (.126 g) was dissolved in n-pentane (10 mL) and placed on a silica gel column (30 g) that had been wet-packed with n-pentane. The column was eluted with n-pentane (500 mL) to remove the saturate hydrocarbons. Aromatic hydrocarbons were eluted from the column using 85% n-pentane–15% benzene (250 mL) and 60% benzene–40% methanol (250 mL). UV analyses of the saturate fraction indicated that trace amounts of aromatic hydrocarbons were present. The amount of saturates in the aromatic fraction, if any, is unknown.

Results and Discussion

A Wilmington, California crude oil was selected for this study because a considerable amount of data is available concerning the compound-type composition of high-boiling distillates (6, 7) and residue (8) from this crude oil. Because methods have been developed for analyzing the high molecular weight polar compounds—compounds suspected to be the building blocks of asphaltenes—separation and analyses of the asphaltenes using the same analytical methods allows a comparison of the composition of the asphaltenes with the high-boiling distillates and residue.

Separation of the Asphaltene. Table I shows the weight percent of the asphaltene fractions and subfractions produced by the separation scheme. The acid fraction, amounting to 81% of the total asphaltene, is the largest fraction isolated by the separation scheme. The primary prerequisite for a compound type to be defined as an acid by the anion, resin appears to be the ability of the compound type to hydrogen bond to the anion resin. Earlier work with distillates and residues identified compound types such as carboxylic acids, phenols, amides, and carbazoles as the major components of an acid fraction (6). Table I shows that Subfraction 3, the subfraction containing the strongest (most readily hydrogen bondable) acids, is more than half of the total acid fraction.

The base fraction represents only 12% of the total asphaltene. Previous work (7) with distillates and residues showed that base fractions isolated by the cation resin contained small amounts of carbazoles and amides; most of the bases were strong bases such as pyridine benzologs

Table I. Weight % of Wilmington Asphaltene Fractions and Subfractions

Sample	Wt %	
Acid fraction	22	
Acid subfraction 1	15	
Acid subfraction 2	44	
Acid subfraction 3		
Total acid fraction		81
Base fraction		
Base subfraction 1	1	
Base subfraction 2	3	
Base subfraction 3	8	
Total base fraction		12
Neutral nitrogen fraction		1
Saturate hydrocarbon fraction		3
Aromatic hydrocarbon fraction		2
Total recovery		99

Table II.   Elemental Analyses of Wilmington Asphaltene and Asphaltene Fractions and Subfractions

Sample	Carbon	Hydrogen	Nitrogen	Sulfur	Oxygen
	Wt %				
Total asphaltene	83.68	8.57	2.18	2.52	2.80
Acid subfraction 1	83.35	8.02	2.37	2.78	3.24
Acid subfraction 2	82.66	8.25	2.64	2.05	4.36
Acid subfraction 3	80.42	8.29	2.03	2.13	6.97
Base subfraction 1	83.41	8.08	1.88	2.75	3.69
Base subfraction 2	79.03	9.23	1.54	6.03	4.19
Base subfraction 3	79.42	8.68	3.20	3.07	5.49
Saturate hydrocarbon fraction	85.80	12.25	0.38	0.91	0.50
Aromatic hydrocarbon fraction	84.52	9.48	0.56	1.83	3.60

and unidentified diaza compounds. Table I shows that base Subfraction 3, the subfraction containing the strongest bases, represents two-thirds of the base fraction.

The neutral nitrogen fraction is 1% of the total asphaltene, and the saturate hydrocarbon and aromatic hydrocarbon fractions are 3% and 2%, respectively. The recovery of material after the separation amounted to 99% of the total asphaltene.

The data in Table I are significant because they suggest that a one-to-one relationship of acids and bases does not exist for petroleum asphaltenes. The precipitation of asphaltenes may be attributed to a phenomenon other than precipitation of acid–base complexes or salts. The data strongly imply that the asphaltenes primarily consist of compounds capable of association through the hydrogen bonding mechanism.

Elemental Analyses. Elemental analyses of the total Wilmington asphaltene, the acid and base subfractions, and the saturate and aromatic hydrocarbon fractions are shown in Table II. In the acid subfractions, nitrogen and sulfur are not concentrated in any one subfraction and the amounts are similar to those of nitrogen and sulfur in the total asphaltene. Large amounts of oxygen are found in all acid subfractions but especially in Subfraction 3, the subfraction expected to contain carboxylic acids. The base subfractions show different distributions of nitrogen, sulfur, and oxygen. Nitrogen is concentrated in base Subfraction 3; sulfur is concentrated in Subfraction 2; and oxygen increases according to subfraction number. Thus, elemental analyses indicate that different compound types are being concentrated in different subfractions. Elemental analyses were not obtained for the neutral nitrogen fraction because the sample was too small. The saturate hydrocarbon fraction contains only small amounts of nitrogen, sulfur, and oxygen. The aromatic hydrocarbon fraction contains small amounts of nitrogen but relatively large amounts of sulfur and oxygen—probably thiophenic sulfur and furan-type oxygen. The amount of nitrogen, oxygen, and sulfur in the total asphaltene was compared with the sum of the nitrogen, oxygen, and sulfur in the fractions and subfractions. The nitrogen and sulfur in the total asphaltene equaled the sum of the amounts found in the separated fractions and subfractions. The analyses for oxygen appear to be in error because more oxygen was found in the fractions and subfractions than was found in the total asphaltene. The trends shown by the oxygen analyses are probably correct, but the actual values are probably in error.

Infrared Spectra. IR spectra of the asphaltene acid and base subfractions are shown in Figures 1 and 2 together with IR spectra of similar subfractions generated from the Wilmington 675°C residue. These spectra (1) demonstrate that chemically meaningful separations have been made by the separation scheme, (2) characterize the compound types in the acid and base subfractions, and (3) show that the compound types in the asphaltene are the same compound types observed in high-boiling distillates and residues.

The partial IR spectrum of acid Subfraction 1 shows IR absorption at 3460 cm^{-1} because of the pyrrolic nitrogen N–H absorption of carbazole-like compounds. Amide carbonyl absorption appears at 1685 cm^{-1}. The partial IR spectrum of acid Subfraction 2 shows the same two IR bands and additional bands at 3585 cm^{-1} and 1650 cm^{-1} owing to phenols and a second amide type. The partial IR spectrum of acid Subfraction 3 shows phenol absorption at 3585 cm^{-1}, pyrrolic nitrogen absorption at 3460 cm^{-1}, and strong carbonyl absorption at 1695 cm^{-1} and 1725 cm^{-1} characteristic of carboxylic acid dimers and monomers. In addition, absorption of hydrogen-bonded carboxylic acid and phenolic hydroxyl groups can be seen in the region of 3500–2300 cm^{-1}.

The partial IR spectrum of base Subfraction 1 shows N–H absorption at 3460 cm^{-1}, amide absorption at 1690 cm^{-1}, and aromatic absorption at 1600 cm^{-1}. Base Subfraction 2 shows increased amounts of aromatic absorption at 1600 cm^{-1} and an additional band at 1720 cm^{-1}, which is thought to be an amide carbonyl absorption. Subfraction 3 shows N–H absorption at 3460 cm^{-1}, amide absorption at 1685 cm^{-1}, and large amounts of aromatic absorption at 1600 cm^{-1} that shows asymmetry typical of pyridine benzologs. Strong bases such as pyridine benzologs appear to be the predominant basic compound type in Subfraction 3.

Molecular Weight Data. Molecular weight data for the total asphaltene and for asphaltene subfractions determined by vapor-pressure osmometry (VPO), low resolution MS, and quantitative IR spectrometry

(6) are shown in Table III. Observed molecular weights are dependent on the method of molecular weight determinations. The data in Table III show two trends: (1) by VPO the acid and base subfractions have higher average molecular weights than the hydrocarbon fractions; and (2) average molecular weights determined by methods other than VPO are lower than those determined by the VPO method. We interpret these data to mean that molecular association of polar molecules is occurring in benzene solvent and the observed VPO weights are aggregate weights and not the molecular weights of individual molecules. The inconsistent values of the VPO weights apparently result from different degrees of association of different compound types rather than large differences in

Table III. Molecular Weight of Total Asphaltene and Asphaltene Fractions and Subfractions

Sample	*Molecular Wt*		
	VPO (Benzene)	*Mass Spectrometry*	*Quantitative IR*
Total asphaltene	2010		
Acid subfraction 1	2160		
Acid subfraction 2	1630		
Acid subfraction 3	1220		584
Base subfraction 1	1490		
Base subfraction 2	1130		
Base subfraction 3	2200	500	
Saturate hydrocarbon fraction	830	630	
Aromatic hydrocarbon fraction	840	500	

actual molecular weight of different compound types. The average weight of individual molecules appears to be in the 500 to 800 range, similar to compounds in the high-boiling distillates and vacuum residue studied previously.

Conclusions from the Data. The preliminary data presented here show that (1) the separation scheme separated asphaltenes according to compound type; (2) the Wilmington asphaltenes are a complex mixture of predominantly polar compound types; (3) acids, as defined by the anion-exchange resin and IR spectrometry, are the predominant compound types in the mixture, amounting to ~ 80% of the asphaltene; and (4) the molecular weights of most individual molecules range from ~ 500 to 800. These data suggest an answer to the question posed earlier: "Why

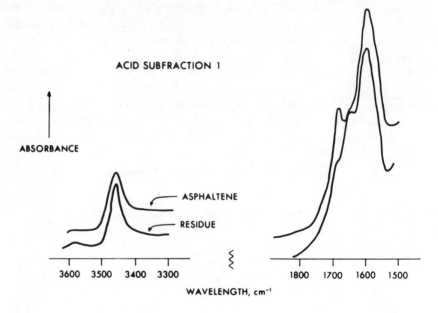

(a)

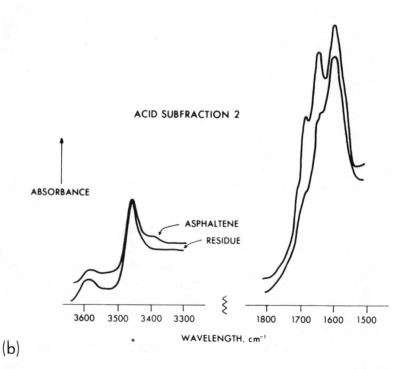

(b)

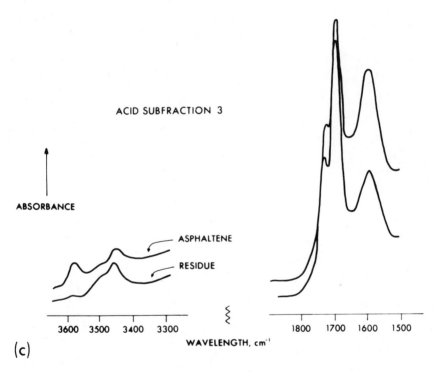

ACID SUBFRACTION 3

ABSORBANCE

ASPHALTENE

RESIDUE

WAVELENGTH, cm⁻¹

(c)

Figure 1.   Infrared spectra

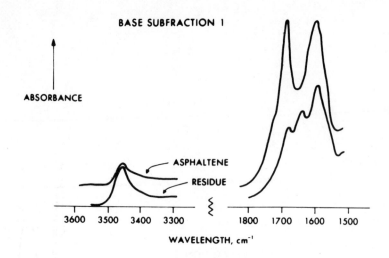

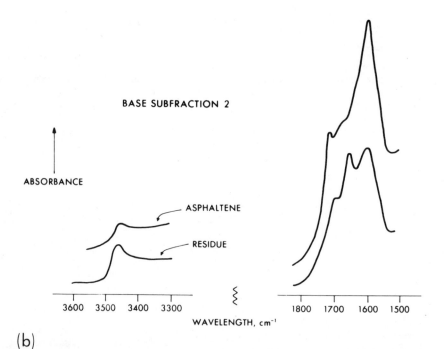

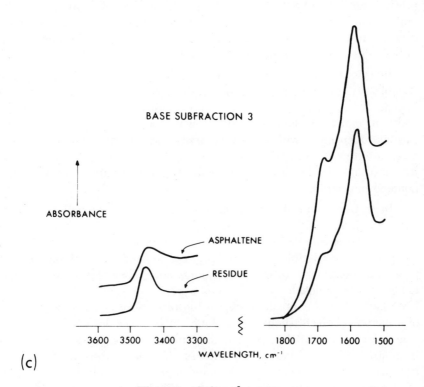

Figure 2. Infrared spectra

are asphaltenes precipitated from solution in petroleum by the addition of a hydrocarbon solvent such as n-pentane?"

Explanation of Asphaltene Formation. The precipitation of asphaltenes is explained in the following manner. Petroleum is a delicately balanced mixture of compounds that depend upon each other for solubility. When the composition is changed, for example by adding large amounts of n-pentane to the oil, the balance is upset and some compounds precipitate. The two factors primarily responsible for maintaining the mutual solubility of the compounds in the complex mixture are the ratio of polar to nonpolar molecules and the ratio of high molecular weight to low molecular weight molecules.

In this discussion, polar compounds are defined as those that are capable of hydrogen bonding with other polar molecules. Thus, carboxylic acids, phenols, carbazoles, and amides are polar molecules. In addition, molecules such as pyridine benzologs are polar because they can hydrogen bond with carboxylic acids and phenols. Nonpolar molecules are those such as normal alkanes, cyclic alkanes, and aromatic hydrocarbons—molecules that normally do not associate with hydrogen-bonding molecules.

In certain circumstances, polar and nonpolar compounds are essentially immiscible. The immiscible nature of water and n-pentane is an example. In a complex mixture such as petroleum, polar and nonpolar compounds are miscible (mutually soluble) as long as a suitable ratio of polar and nonpolar molecules is maintained. When this ratio is maintained, polar molecules dissolve other polar molecules (like dissolves like) and solution is possible. When the ratio is altered by the addition of a nonpolar solvent such as n-pentane, polar molecules are less soluble. The polar molecules then form hydrogen-bonded aggregates of nonuniform size and precipitate as asphaltenes. It is not surprising that acids such as carboxylic acids, phenols, amides, and carbazoles represent the most abundant compound types in the Wilmington asphaltene because they are the compounds most incompatible with n-pentane by virtue of being the most polar (hydrogen bondable) molecules in the petroleum. Pyridine benzologs are another polar compound type that are incompatible with a nonpolar solvent and also precipitate. The occurrence of the nonpolar compound types (saturate hydrocarbons and aromatic hydrocarbons) with the polar components of the asphaltene may result from occlusion of nonpolar molecules with aggregates of polar molecules.

The ratio of low molecular weight compounds to high molecular weight compounds is another factor in maintaining solubility of all compounds in petroleum. When this ratio is upset, large molecules precipitate (form asphaltene). This study shows that the precipitated asphaltenes have an average molecular weight of ~ 500 to 800; whereas the average molecular weight of the total crude oil is estimated to be 200. In principle, the solubility of molecules in a mixture is an additive phenomenon—a C_{10} molecule depends upon a C_5 molecule for solubility, a C_{15} molecule depends upon both a C_{10} and C_5 molecule for solubility, a C_{20} molecule depends upon a C_{15}, C_{10}, and C_5 molecule for solubility, and so on. In a solution containing very large molecules, such as C_{60} molecules, the ratio of low, medium, and high molecular weight compounds is a delicate balance and addition of small molecules such as n-pentane will destroy the balance and cause precipitation of the largest molecules. When n-pentane is added to petroleum, the petroleum changes from a solution having an average molecular weight of, for example, 200 to a solution having an average molecular weight nearer that of n-pentane, 72. A solution having a small average molecular weight cannot dissolve the largest molecules in petroleum, and so these molecules precipitate as asphaltenes. The amounts of asphaltenes precipitated from petroleum by heptane, pentane, and propane are different because the average molecular weight of the petroleum is different when these solvents are added to the petroleum. Of these solvents, propane causes the largest amount of asphaltene formation because the average molecular weight of the petroleum is lowered more than when the other solvents are used. The high molecular weight molecules are less soluble in a propane-like solvent than in a pentane- or heptane-like solvent.

Although this chapter is limited to experimental work on petroleum asphaltenes, we suggest that asphaltene precipitation is a phenomenon common to complex organic solutions in general. Thus, the principles that apply to the formation of petroleum asphaltenes also apply to coal liquids, shale oil, tar sand bitumen, or any complex solution of organic compounds. The composition of the asphaltenes from different organic solutions differs because the compositions of the solutions are different. In general, the most polar components of a mixture would be expected to precipitate when nonpolar solvents are used to generate the asphaltene. For example, work on coal liquids (9, 10) shows that asphaltenes from coal liquids contain large amounts of phenols. In coal liquids, phenols represent one of the most polar compound types and thus precipitate first from solution when pentane is added. The amount of material precipitated depends on the ratio of polar to nonpolar compounds in the solution and on the ratio of low molecular weight to high molecular weight compounds in the solution. The amount of asphaltene precipitated will depend entirely on how much the delicate balance of the solution is upset by the addition of a particular solvent.

Summary

An asphaltene was precipitated from a Wilmington, California crude oil by adding *n*-pentane to the crude oil. The asphaltene was separated and analyzed according to compound type, using the analytical technique developed previously for the analysis of high-boiling distillates and residues. The asphaltene contains some of the same compound types found in high-boiling distillates and residues. High molecular weight (averaging 500 to 800) polar compounds such as carboxylic acids, phenols, amides, carbazoles, and pyridine benzologs represent the major components in this asphaltene.

We suggest that the composition of an asphaltene is dependent upon the composition of the complex organic mixture from which it is generated, the ratio of polar to nonpolar compounds, the ratio of low molecular weight to high molecular weight molecules, and the solvent used to precipitate the asphaltene. Therefore, the composition of any asphaltene is generally predictable if the composition of the complex organic mixture from which it is generated is known. Asphaltenes precipitated by low molecular weight hydrocarbon solvents will include the most polar species and the highest molecular weight species present in the complex organic mixture. Molecular weights determined by measurements made in solution such as VPO are actually aggregate weights and depend on how the molecule–aggregate equilibrium is affected by a particular solvent.

Literature Cited

1. Helm, R. V., Latham, D. R., Ferrin, C. R., Ball, J. S., *Chem. Eng. Data Ser.* (1957) **2**(1), 95.
2. Ferris, S. W., Black, E. P., Clelland, J. B., *Prepr., Div. Pet. Chem., Am. Chem. Soc.* (1966) **11**(2), B-130.
3. Middleton, W. R., *Anal. Chem.* (1967) **39**, 1839.
4. Altgelt, K. H., *Prepr., Div. Pet. Chem., Am. Chem. Soc.* (1965) **10**, 29.
5. Dickie, J. P., Yen, T. F., *Anal. Chem.* (1967) **39**, 1847.
6. McKay, J. F., Cogswell, T. E., Weber, J. H., Latham, D. R., *Fuel* (1975) **54**, 50.
7. McKay, J. F., Weber, J. H., Latham, D. R., *Anal. Chem.* (1976) **48**, 891.
8. McKay, J. F., Amend, P. J., Harnsberger, P. M., Cogswell, T. E., Latham, D. R., *Am. Chem. Soc., Div. Fuel Chem., Prepr.* (1976) **21**(7), 52.
9. Sternberg, H. S., *Am. Chem. Soc., Div. Fuel Chem., Prepr.* (1976) **21**(7), 1.
10. Schweighardt, F. K., Retcofsky, H. L., Raymond, R., *Am. Chem. Soc., Div. Fuel Chem., Prepr.* (1976) **21**(7), 27.

RECEIVED August 5, 1977.

Organometallic Complexes in Domestic Tar Sands

J. F. BRANTHAVER and S. M. DORRENCE

U. S. Department of Energy, Laramie Energy Research Center, Laramie, WY 58102

The vanadium, nickel, and porphyrin contents of organic materials derived from a number of domestic tar sands have been determined. Some of the tar sands were extracted in a Soxhlet extractor, while others were extracted at room temperature by a series of solvents of increasing polarity beginning with n-heptane and ending with pyridine. Most of the samples were from outcrops of various Utah deposits, although some core samples were investigated. Porphyrins were found in some of the outcrops and cores. With one exception, materials having a high sulfur content also have a ratio of vanadium to nickel of greater than unity, while low-sulfur materials have vanadium-to-nickel ratios of less than unity.

Domestic tar sands are known to contain over 30 billion barrels of oil, most of which occurs in a few deposits in Utah (1). The Laramie Energy Research Center (LERC) is investigating many of these deposits in order to determine the potential for exploitation of this resource.

One of the properties of fossil fuels of interest to refiners is metal content. Vanadium in particular is troublesome in refinery operations (2). In some crude oils it occurs in concentrations up to 1000 ppm (3). Of the other metals in petroleum, only nickel is known to be present in concentrations approaching that of vanadium. These two metals are complexed with porphyrins and other ligands, the exact nature of which is unknown (4). Crude oils that are high in sulfur are usually high in vanadium (5), and it is interesting to determine whether or not this criterion also applies to tar sands.

Experimental

Some of the tar sands were extracted exhaustively with benzene–ethanol (3:1) in a Soxhlet extractor to yield organic materials, and only those materials so obtained will be subsequently referred to as bitumens. This process does not remove all organic materials from some tar sands, and so another method of extraction was used. In the alternate procedure, tar sands first were pulverized in a mortar and pestle. Then the sample was transferred to Erlenmeyer flasks and treated with successive volumes of *n*-heptane until less than 25 mg of material was dissolved. Each volume of heptane solution was filtered on a sintered glass funnel (coarse porosity), and the heptane then was removed. After heptane was observed to dissolve little material, the solvent was changed to benzene and after this to a mixture of benzene–methanol (1:1). Finally, the sands were extracted with several portions of warm pyridine.

Vanadium and nickel concentrations were determined colorimetrically by forming tungstovanadic acid and nickel dimethylglyoxime according to the procedure of Bean (3). Measurements were performed using a Beckman Acta M-IV spectrophotometer. Porphyrin analyses were done by the direct intergral technique of Bean and Sugihara (6) using a Beckman DK-2 spectrophotometer. Analyses for sulfur and nitrogen were carried out by the Technical Services Section of Laramie Energy Research Center.

Asphaltenes were obtained by contacting bitumens with large volumes of heptane.

Results and Discussion

Compared with a high-sulfur, metal-rich crude oil from the Boscan field of Venezuela (3), the vanadium contents of most of the tar sand bitumens (*see* Table I) are relatively low, particularly in view of the high sulfur values of some of the bitumens. If tar sand bitumens are the residues of crude oils which have lost their light ends, it might be expected that metals, which are always concentrated in the heavier

Table I.   Analyses of Various Tar Sand Bitumens

	$\mu moles/g$		V/Ni	Vanadyl Porphyrin, $\mu moles/g$	%	
	V	Ni			S	N
Boscan, Venezuela, crude oil	22	2.0	11	10	—	—
Battle Creek, WY	2.8	0.8	3.5	0.5	5.7	0.6
Asphalt Ridge, UT	0.7	1.1	0.6	< 0.1	0.4	0.8
P.R. Spring, UT	1.1	1.8	0.6	< 0.1	0.6	1.3
Tar Sand Triangle, UT	3.8	1.2	3.2	< 0.1	4.5	0.6
Edna, CA	1.9	2.5	0.8	< 0.1	3.0	1.2
Sant Rosa, NM	0.5	0.4	1.2	< 0.1	2.3	0.2
Athabasca, Alta.	4.9	1.5	3.3	0.9	4.7	0.4

Table II. Analyses of Solvent-Extracted Fractions from Uinta Basin, Utah Tar Sands

		μmoles/g			Wt %	
	Wt, g	V	Ni	V/Ni	S	N
Lake Fork, outcrop (244 g)						
heptane-soluble fraction	26.1	0.2	0.4	0.50	0.4	1.0
benzene-soluble fraction	3.7	2.7	1.9	1.42	0.9	1.9
combined heptane- and benzene-soluble fractions	29.8	0.5	0.6	0.83	0.5	1.1
Whiterocks, outcrop (233 g)						
heptane-soluble fraction[a]	18.9	0.3	0.8	0.37	0.3	1.4
benzene-soluble fraction	0.3	4.5	3.9	1.15		
combined heptane- and benzene-soluble fractions	19.2	0.4	0.8	0.50		
S.E. Asphalt Ridge, outcrop (281 g)						
heptane-soluble fraction[a]	28.7	0.5	1.5	0.33	0.2	1.2
benzene-soluble fraction	1.4	8.6	5.4	1.59	0.8	1.9
combined heptane- and benzene-soluble fractions	30.1	0.9	1.7	0.53	0.2	1.2
Raven Ridge, outcrop (400 g)						
heptane-soluble fraction	7.3	0.6	0.5	1.2	0.6	1.0
benzene-soluble fraction	7.9	0.9	3.0	0.3	0.6	1.8
combined heptane- and benzene-soluble fractions	15.2	0.8	1.8	0.44	0.6	1.4
P.R. Spring, outcrop (254 g)						
heptane-soluble fraction[a]	15.2	0.4	1.6	0.25	0.6	1.7
benzene-soluble fraction	2.1	4.9	5.2	0.94	0.6	2.4
combined heptane- and benzene-soluble fractions	17.3	0.9	2.0	0.45	0.6	1.8
Sunnyside, outcrop (249 g)						
heptane-soluble fraction	23.0	0.6	0.7	0.86	0.9	1.0
benzene-soluble fraction	4.8	2.1	3.5	0.60	0.2	1.6
combined heptane- and benzene-soluble fractions	27.8	0.9	1.2	0.75	0.8	1.1
Argyle Canyon, outcrop (328 g)						
heptane-soluble fraction	10.2	0.3	0.3	1.00	1.3	0.6
benzene-soluble fraction	2.8	2.3	3.2	0.72	1.4	1.2
combined heptane- and benzene-soluble fractions	13.0	0.7	0.9	0.77	1.3	0.7

[a] Nickel porphyrins were detected in these samples.

fractions of petroleum, would be present in greater amounts than are observed in these bitumens.

Radchenko (5) has indicated that for a large number of crude oils high-sulfur content correlates with high-vanadium content, high vanadyl porphyrin content, and high ratios of vanadium to nickel. Of the high-sulfur bitumens examined in the present study, only the Wyoming and Athabasca bitumens contained significant amounts of vanadyl porphyrins. These two bitumens and two of the other high-sulfur samples have high ratios of vanadium to nickel, but the Edna bitumen, which is also high in sulfur, has a low ratio of vanadium to nickel. The high-sulfur Santa Rosa bitumen is the lowest of the seven in vanadium content. The two samples from the Uinta basin of Utah (Asphalt Ridge and P. R. Spring) are low in sulfur and, as expected, contain small amounts of vanadium and have low ratios of vanadium to nickel.

The Battle Creek, Wyoming outcrop sample has substantial amounts of vanadyl porphyrins, indicating that petroporphyrins may be preserved even after prolonged exposure to weathering agents.

Table II shows the results obtained by extracting several Uinta Basin, Utah outcrops with successive organic solvents. All outcrop samples are fairly low in sulfur, most are quite high in nitrogen, and all have low ratios of vanadium to nickel. Only the Raven Ridge sample, which was collected in a creek bed, has a very large fraction of organic material that is not soluble in heptane: Benzene–methanol (1:1) and pyridine did not extract much material from any of these samples, so analytical data from these materials are not included in the table. The asphaltenes extracted from P. R. Spring and Southeast Asphalt Ridge tar sands are quite rich in nickel ($5 \mu mol/g$), and nickel porphyrins are found in the heptane-soluble fractions of these tar sands as well as is the heptane-soluble fraction of Whiterocks tar sands. Crudes derived from nonmarine sources are usually much higher in nickel content than in vanadium content, and the Uinta Basin tar sands deposits are all of lacrustine origin and are of tertiary age.

The organic materials derived from Tar Sand Triangle and Circle Cliffs outcrop samples, which lie east and west of Utah's Henry Mountains (200 miles south of the Uinta Basin), contrast markedly with those of the Uinta Basin (Table III). In each case, the South Utah heptane-soluble materials are a clear amber color and contain no detectable amounts of nickel. Simple chromatographic separations yield substantial amounts of colorless, high–boiling liquids which are free from nitrogen but still contain 1–2% sulfur. The asphaltene fractions are all black solids, and that portion which can only be recovered by pyridine is quite large. The organic materials extracted from these outcrops are all high in sulfur, have vanadium to nickel ratios ranging from 2.9 to 4.3, and do not contain porphyrins.

These deposits are the largest known tar sand, or perhaps more appropriately, bituminous sandstone deposits in the United States. The presence of such a large percentage of organic material that can only be

Table III. Analyses of Solvent-Extracted Fractions
from Southern Utah Tar Sands

		$\mu moles/g$			Wt %	
	Wt, g	V	Ni	V/Ni	S	N
Elaterite Basin, outcrop (246 g)						
heptane-soluble fraction	12.6	0.4	0.0	>20	3.3	0.5
benzene-soluble fraction	9.8	13.7	2.8	4.89	5.3	1.1
benzene–methanol-soluble fraction	0.8	5.8	2.3	2.52	6.4	1.1
pyridine-soluble fraction	2.5	9.5	3.5	2.71	4.4	1.1
combination of all fractions	25.7	6.5	1.5	4.33	4.2	0.8
Teapot Rock, outcrop (392 g)						
heptane-soluble fraction	2.4	0.5	0.0	>20	1.7	0.6
benzene-soluble fraction	1.9	3.2	0.9	3.55	4.8	0.7
benzene–methanol-soluble fraction	2.5	4.0	2.5	1.60	4.1	0.8
pyridine-soluble fraction	4.6	7.7	2.3	3.36	5.0	1.9
combination of all fractions	11.4	4.6	1.6	2.87	4.0	1.2
Tar Cliffs, outcrop (278 g)						
heptane-soluble fraction	4.3	0.5	0.0	>20	3.0	0.6
benzene-soluble fraction	3.8	4.5	1.6	2.81	4.6	0.7
benzene–methanol-soluble fraction	1.3	3.8	2.4	1.58	4.6	1.1
pyridine-soluble fraction	2.9	6.9	2.9	2.37	5.0	1.0
combination of all fractions	12.3	4.0	1.3	3.07	4.1	0.8
Circle Cliffs, outcrop (317 g)						
heptane-soluble fraction	5.8	0.8	0.0	>20	3.4	0.2
benzene-soluble fraction	5.4	4.8	1.2	4.00	5.9	0.8
benzene–methanol-soluble fraction	1.8	6.4	3.5	1.82	4.9	1.2
pyridine-soluble fraction	5.5	9.8	3.9	2.51	5.6	1.6
combination of all fractions	18.5	5.2	1.8	2.89	4.9	0.9

solvent-extracted by solvents such as pyridine undoubtedly will affect the choice of method used in their commercial exploitation.

Core samples from the Family Butte deposit in Utah's San Rafael Swell (midway between the Uinta Basin and the Tar Sand Triangle) were examined also (Table IV). Samples from two producing zones separated by strata clean of organic materials were chosen for investigation. One zone of 161–169-ft depth and the other of 324–333-ft depth yielded similar organic materials upon solvent extraction of core samples. The heptane-soluble fraction constitutes ~ 60% of the total of extracted materials in each case, and both are dark brown in color. The asphaltene fractions are very high in sulfur and nitrogen. Only small amounts of materials were recovered from the sands by means of benzene–methanol and

Table IV.   Analyses of Family Butte, Utah Tar Sands

	Wt, g	$\mu moles/g$ V	$\mu moles/g$ Ni	V/Ni	Wt % S	Wt % N
Family Butte, core, 161-169 ft depth (308 g)						
heptane-soluble fraction[a]	9.5	0.5	0.1	5.00	4.6	0.5
benzene-soluble fraction[a]	5.4	8.7	2.5	3.48	6.5	1.1
benzene–methanol-soluble fraction	0.3	19.9	11.2	1.78	9.0	1.0
pyridine-soluble fraction	0.3	12.0	3.7	3.24	6.0	2.1
combination of all fractions	15.5	3.9	1.2	3.25	5.4	0.8
Family Butte, core, 324-333 ft depth (263 g)						
heptane-soluble fraction[a]	5.3	0.5	0.1	5.00	4.3	0.4
benzene-soluble fraction[a]	3.0	7.5	2.6	2.88	9.5	1.3
benzene–methanol-soluble fraction	0.2	6.4	1.8	3.55	5.1	1.9
pyridine-soluble fraction	0.2	11.0	3.6	3.06	5.6	1.8
combination of all fractions	8.7	3.3	1.1	3.00	6.1	0.8

[a] Vanadyl porphyrins were detected in these fractions.

pyridine extraction. Organic materials from both zones have similar vanadium-to-nickel ratios and contain trace amounts of vanadyl porphyrins, suggesting that each zone contains organic material from the same source.

Table V shows metal distributions in two other domestic tar sand fractions. In the Battle Creek bitumen from Wyoming, bitumen metals and porphyrins are concentrated in the asphaltenes, as expected. Fractions obtained from the Edna, California outcrop have vanadium-to-

Table V.   Analyses of Fractions from Edna, California, and Battle Creek, Wyoming, Tar Sands

	$\mu moles/g$ V	$\mu moles/g$ Ni	V/Ni	Vanadyl Porphyrin $\mu moles/g$
Edna, outcrop				
heptane-soluble fraction	0.2	0.9	0.22	< 0.1
benzene-soluble fraction	2.8	5.2	0.54	0.1
benzene–methanol-soluble fraction	3.7	3.7	1.00	< 0.1
pyridine-soluble fraction	3.9	4.0	0.98	< 0.1
Battle Creek, outcrop, bitumen				
petrolenes	1.0	0.2	5.00	0.2
asphaltenes	7.5	2.8	2.68	1.3

nickel ratios of unity or less; yet spectroscopic examinations indicate the presence of vanadyl porphyrins only (chromatographic treatment of the benzene-soluble fraction might, of course, yield some nickel porphyrins).

Conclusions

The organic materials extracted from a number of domestic tar sands contain various amounts of vanadium and nickel, sometimes in substantial quantities. Metalloporphyrin complexes of these two metals have been detected in some outcrop and shallow core samples. If these porphyrins are indigenous petroporphyrins, it is evident that these compounds can survive weathering processes.

The organic matter in these tar sands resembles most crude oils in several respects. Metals always are concentrated in asphaltene fractions. Extracts from tar sands which are high in sulfur content usually have ratios of vanadium-to-nickel greater than unity, while those which are low in sulfur tend to have vanadium-to-nickel ratios less than unity. An exception to this generalization is the Edna, California tar sands, which yield extracts that are high in sulfur content but have vanadium-to-nickel ratios of less than unity.

Literature Cited

1. "Energy from U.S. and Canadian Tar Sands: Technical, Environmental, Economic, Legislative, and Policy Aspects," Report prepared for the Subcommittee on Energy of the Committee on Science and Astronautics, U.S. House of Representatives, 93rd Congress, 2nd Session, U.S. Government Printing Office, Washington, D. C. (1974) p. 9.
2. Farrar, G. L., *Oil Gas J.* (1952) **Apr. 7,** 79.
3. Bean, R. M., Ph.D. Thesis, University of Utah (1961) 83–87.
4. Dickson, F. E., Kunesh, C. J., McGinnis, F. L., Petrakis, L., *Anal. Chem.* (1972) **44,** 978.
5. Radchenko, O. A., "Geochemical Regularities in the Distribution of the Oil-Bearing Regions of the World," Leningrad, 1965; Israel Program for Scientific Translations, Jerusalem, 1968, pp. 200–206.
6. Sugihara, J. M., Bean, R. M., *J. Chem. Eng. Data* (1962) **7,** 269.

RECEIVED October 17, 1977.

11

Analyses of Oil Produced during in Situ Reverse Combustion of a Utah Tar Sand

S. M. DORRENCE, K. P. THOMAS, J. F. BRANTHAVER, and R. V. BARBOUR

Department of Energy, Laramie Energy Research Center, Laramie, WY 82071

Oils produced during an in situ reverse combustion experiment in the Northwest Asphalt Ridge Utah tar sand deposit have been characterized. Elemental analyses, distillation data, pour points, specific gravities, chromatographic separations, IR, and C-13 NMR spectra, and molecular weights are reported for the oils. These data are compared with data obtained on a bitumen sample extracted from a core taken at the site prior to the experiment. Data indicate that most of the oil produced during the experiment is very similar to the original bitumen.

Although considerable effort has been devoted to research and development of the tar sands of Alberta, Canada, much less work has been devoted to tar sands within the United States. Domestic deposits are widespread, but the greatest known resource occurs in Utah. Chemical composition data of Utah bitumens have been the topic of several publications (1–8); however, only a few have given data relative to products resulting from processing of Utah tar sands bitumens (1,7,8). Bunger, Mori, and Oblad recently described analyses of products obtained following thermal cracking of solvent-extracted Utah bitumens (7). Barbour et al. published data relative to analyses of pyrolysis products from whole tar sands (8).

Late in 1975, personnel from Laramie Energy Research Center (LERC), then of the U.S. Energy Research and Development Administration, conducted an in situ recovery experiment in the Northwest Asphalt Ridge deposit near Vernal, Utah. Reverse combustion was chosen as a recovery method because it was felt that this method was more likely to succeed for in situ recovery of tar sands hydrocarbons.

During reverse combustion the hydrocarbons produced pass through heated sand and are less likely to cause plugging problems than in a forward combustion method in which products or bitumen can migrate into unheated sand.

The field experiment, which has been described elsewhere (*9, 10, 11*), produced liquid organic, aqueous, and gaseous materials. This chapter describes the characterization of product oils from the experiment, and these data are compared with data obtained on bitumen extracted from core samples taken at the field site prior to the burn.

Experimental

Experimental Site. The experiment was conducted in the Northwest Asphalt Ridge tar sand deposit, about 5 miles west of Vernal, Utah. The site was on property owned by Sohio Oil Co., which granted use through a cooperative agreement. The test zone was ~ 300 ft below the surface and averaged ca. 10 ft in thickness. The zone was in the Rim Rock sandstone member of the Mesa Verde formation and apparently was isolated by impermeable layers above and below the selected interval. Ignition was accomplished on November 25, 1975 and the experiment was continued until December 19. During this time, oil production was 65 barrels. Details on cores from the site, control equipment, production equipment, on-site tests prior to ignition, and conduct of the experiment have been presented elsewhere (*9, 10, 11*).

Collection of Oil Samples. In the production train, two separators were utilized in series. The first separator condensed higher boiling oil components, while the second separator condensed lower-boiling oils and water. For reference purposes, the product oils from the first and second separators are referred to as heavy and light oils, respectively.

Preparation of Oils for Analysis. The light oils were used as sampled. All were essentially free from water and particulates. The heavy oils contained some water and sand. Water was removed by using a Barrett receiver while azeotroping with toluene. The solutions of heavy oils in hot toluene subsequently were filtered through glass-fritted filter funnels with fine pore sizes (4–5.5 μ) to remove sand particles. Then the sand was washed several times with hot toluene. Toluene was removed in vacuo using a rotary film evaporator and a hot water bath. The final removal of solvent was accomplished at 4 mm of mercury pressure and at 90°C waterbath temperature.

Bitumen from the core samples was extracted with boiling toluene, using a modified Soxhlet extractor (*12*), and the solution was filtered and evaporated as described for heavy oils.

Analytical Methods. Carbon, hydrogen, nitrogen, and sulfur analyses and Hempel distillations (*13*) were performed using conventional analytical techniques. Nickel and vanadium contents were determined by the methods of Bean (*14*). These methods involve the colorimetric determination of nickel dimethylglyoxime and tungstovanadic acid, respectively. Simulated distillations were done by the method of Poulson et al. (15). IR spectra were obtained using a Perkin–Elmer Model 621 IR spectrophotometer. Using a Varian CFT–20 spectrometer, ^{13}C NMR spectra were recorded using tetramethylsilane as an internal standard. Molecular weights were determined by vapor-phase osmometry using three concentrations of sample.

The oils and bitumen were separated chromatographically on silica gel into saturates, aromatics, polar aromatics, and asphaltenes fractions by the SAPA method of Barbour et al. (8).

Results and Discussion

The light oils produced during the field experiment are very different from the heavy oils. In this section, the two types of product oils will be compared with bitumen separately. Both the light and heavy oils sampled on the fourth day of the run were quite different from other samples in their own category. These differences will be summarized in the last paragraph of each subsection.

Comparison of Light Oils with Bitumen. Elemental analyses are presented in Table I. From these data, hydrogen-to-carbon atomic ratios were calculated. This ratio for light oils is in the range of 1.83–1.90, indicating a high degree of saturation. The bitumen has a ratio of 1.64. The light oils contain much less nitrogen and sulfur than the bitumen does. Nickel and vanadium contents are considerably lower for light oils than for bitumen.

Some physical properties and some distillation data for oils and bitumen are shown in Table II. Specific gravities and pour points are lower for light oils than for bitumen. Viscosities (ca. 6 cP at 77°F) for light oils are much less than the viscosity of the bitumen (ca. 3,000,000 cP at 77°F). Distillation data show that the light oils contain much more material boiling below the two cut points (275° and 425°C) than

Table I. Elemental Analysis of Bitumen and Product Oils

	Wt %				ppm	
Sample (day)	Car-bon	Hydro-gen	Nitro-gen	Sul-fur	Vana-dium	Nickel
Bitumen	85.3	11.7	1.02	0.59	115	80
Light oil (4th)	85.6	11.6	0.32	0.66	44	4
Light oil (10th)	84.8	13.0	0.13	0.21	< 1	< 1
Light oil (22nd)	85.4	13.3	0.12	0.17	10	5
Light oil (24th)	85.1	13.6	0.03	0.17	20	5
Heavy oil (4th)	85.7	12.0	1.32	0.57	18	< 1
Heavy oil (11th)	85.1	11.8	0.90	0.40	18	59
Heavy oil (21st)	85.3	12.0	1.09	0.35	18	76
Heavy oil (24th)	85.0	12.1	0.89	0.62	150	65

Table II.   Physical Properties and Distillation Data for Bitumen and Product Oils

Sample (day)	Specific Gravity (60°F/60°F)	Pour Point °F	% Distillable[a] 275°C	% Distillable[b] 425°C
Bitumen	0.97	150	6	19.5
Light oil (4th)	0.92	10	56	80
Light oil (10th)	0.87	< −5	78	99
Light oil (22nd)	0.84	< −5	92	96
Light oil (24th)	0.85	< −5	80	88
Heavy oil (4th)	1.04	185	28	56
Heavy oil (11th)	0.96	180	5	28
Heavy oil (21st)	0.98	180	8	26
Heavy oil (24th)	1.00	185	18	30

[a] Material boiling below 275°C is in the boiling range of gasoline, naphtha, and kerosine.

[b] The end point for the Hempel distillation is equivalent to 425°C. Data for bitumen and heavy oils were obtained by simulated distillation, and data for light oils were obtained by Hempel distillation.

the bitumen contains. In the Hempel procedure (*13*), 275°C is the temperature at which the distillation at 760-mm mercury pressure is discontinued. The distillation is continued at 40-mm mercury pressure to provide an endpoint equivalent to 425°C at 760-mm mercury pressure.

Results of the application of the SAPA chromatographic procedure are given in Table III. The light oils contain mostly saturate fractions. The bitumen contains 48% saturate fraction. The aromatic, polar aromatic, and asphaltene fractions of light oils are reduced greatly compared with bitumen. These data indicate that the light oils could be composed mainly of the lower boiling components in the bitumen or are cracked products from the bitumen. A combination of the two alternatives is also possible.

IR spectra of light oils indicate a considerable decrease in aromatic content compared with bitumen. The light oils have strong IR absorptions for olefins at ~ 1640 cm^{-1} and between 840 and 100 cm^{-1}. The bitumen has no noticeable olefin content. These data indicate that some cracked products were formed during recovery. The decrease in aromatic content could result from aromatic components being higher boiling in the original bitumen or from aromatic components forming coke during processing.

To characterize the light oils more completely, ^{13}C NMR were obtained (*see* Table IV). Using a combination of IR and NMR data, several conclusions can be made about the composition of bitumen and produced oils. Because IR spectra indicate that the bitumen has unde-

Table III. Chromatographic Separation Data for Bitumen and Product Oils

Wt %

Sample (day)	Saturates	Aromatics	Polar Aromatics	Asphaltenes
Bitumen	48.1	18.4	27.2	6.3
Light oil (4th)	74.4	14.4	11.2	< 0.1
Light oil (10th)	86.7	8.6	4.6	0.1
Light oil (22nd)	92.3	5.9	1.8	< 0.1
Light oil (24th)	91.3	5.7	3.0	< 0.1
Heavy oil (4th)	44.9	19.3	12.6	23.1
Heavy oil (11th)	42.1	26.3	13.8	17.7
Heavy oil (21st)	47.1	24.1	6.0	22.8
Heavy oil (24th)	42.9	12.0	10.7	34.4

tectable olefin content, the carbon atoms in bitumen with resonance in the range between 110- and 150-ppm shift from tetramethylsilane are assumed to be aromatic. The range between 5- and 60-ppm shift are attributed to saturate–alicyclic carbon atoms. Integration of areas under the peaks indicate that the bitumen contains ca. 13% aromatic and 87% saturate–alicyclic carbon atoms.

IR spectra of light oils indicate the presence of olefins and that aromatic contents are low. Apparently the olefinic carbon atoms in the produced oils are in the same range (between 110 and 150 ppm) as aromatic carbon atoms in the ^{13}C NMR spectra because no differentiation is possible by observing these spectra. Of aromatic and olefinic types, based upon IR data, the light oils are proposed to contain mostly olefinic

Table IV. C-13 NMR Data for Bitumen and Product Oils

Sample (day)	Carbon Type, % [a]	
	Aromatic–Olefinic	Saturate–Alicyclic
Bitumen	13	87
Light oil (4th)	19	81
Light oil (10th)	9	91
Light oil (22nd)	8	92
Light oil (24th)	7	93
Heavy oil (4th)	35	65
Heavy oil (21st)	11	89
Heavy oil (24th)	10	90

[a] Aromatic–olefinic integrals were measured in the 110–150-ppm shift region. Saturate–alicyclic integrals were measured in the 5–60-ppm shift region.

carbon atoms. However, 91–93% of the total carbon atoms in light oils are of the saturate–alicyclic types. Only 7–9% are olefinic or aromatic.

Molecular weights were determined for the light oils using vapor pressure osmometry (VPO) in two solvents, benzene and tetrahydrofuran (THF). THF was used to compare molecular weight values obtained using a more polar solvent with values obtained in benzene. Data are presented in Table V. The light oils have considerably lower molecular weights in either solvent than the original bitumen. The light oils have relatively constant molecular weights in either solvent with lower values in THF indicating some hydrogen bonding of light oils in benzene.

The light oil collected 4 days after ignition is different from the other light oils. It contains less materials boiling below 275°C and less material boiling below 425°C. The hydrogen-to-carbon atomic ratio of

Table V. Molecular Weight Determinations for Bitumen and Product Oils

Sample (day)	Mol Wt in Benzene[a]	Mol Wt in THF
Bitumen	732	683
Light oil (4th)	226	218
Light oil (10th)	245	203
Light oil (22nd)	252	187
Light oil (24th)	254	202
Heavy oil (4th)	475	298
Heavy oil (21st)	510	537
Heavy oil (24th)	654	508

[a] The reported values are the average of three determinations by VPO.

1.61 is similar to the value for the bitumen. About 19% of the carbon atoms are olefinic or aromatic and 81% are saturate or alicyclic as indicated by NMR. This sample contains more sulfur and nitrogen than the other light oils. The specific gravity is 0.92, the pour point is 10°F, and the viscosity is 8 cP at 77°F. Molecular weights do not indicate that this light oil differs greatly from the other light oils.

Comparison of Heavy Oils with Bitumen. Many of the analyses indicate that the heavy oils are very similar to the original bitumen. Hydrogen-to-carbon atomic ratios are in the range of 1.65–1.69, which is similar to the bitumen (1.64). Nitrogen and sulfur contents are similar to values for the bitumen (*see* Table I). Nickel concentrations of heavy oils are slightly lower than the concentration in the bitumen. Vanadium analyses for heavy oils range from lower than to higher than vanadium in the bitumen.

Specific gravities and pour points are higher for the heavy oils than for the bitumen (Table II). Viscosities of heavy oils are too high to be measured at 77°F. Distillation data show that heavy oils contain slightly more material that distills below 425°C than the bitumen contains.

SAPA analyses for heavy oils are variable in that the aromatic fractions range from being less than to greater than values for the bitumen (Table III). Saturates fractions are slightly lower than the bitumen value. Polar aromatics fractions are reduced considerably from the bitumen, but asphaltenes contents are increased greatly. Slight changes in components from polar aromatics fractions could readily result in their insolubility in n-pentane resulting in increased asphaltenes.

The IR spectra reveal evidence for the presence of olefins in the heavy oils; however, the concentration is low. Other aspects of the IR spectra of heavy oils are nearly identical to the spectrum for the bitumen. One slight difference is noticeable in the 1700-cm^{-1} region in that carbonyl absorption is present, indicating some oxidation; however, the extent is not great.

^{13}C NMR examination of heavy oils indicates that they are very similar to the bitumen. Data are summarized in Table IV. Most of the major peaks for the heavy oil are at the same chemical shifts and in the same relative intensities as are observed for the bitumen. Heavy oils contain 10–11% aromatic and/or olefinic carbon atoms, with the remainder being in saturate or alicyclic structures.

Molecular weight data shown in Table V indicate that some modification of bitumen has occurred during processing to produce heavy oils. Values for heavy oils are silghtly lower than the bitumen molecular weights in either benzene or THF. The molecular weights for heavy oils are much higher than molecular weights for light oils.

Much of the data obtained with heavy oil samples indicate that these samples are similar to original bitumen. Data reported elsewhere (11) about this field experiment show that most of the formation did not experience high temperatures. The highest thermocouple reading observed was 350°F, with most readings ca. 220°F. This is sufficient temperature to mobilize the bitumen but is not hot enough to cause extensive thermal cracking and/or distillation of bitumen components.

The heavy oil collected on the fourth day contains more low-boiling components than the other heavy oils. Vanadium and nickel analyses are lower than for bitumen. The hydrogen-to-carbon ratio and nitrogen and sulfur contents are similar to values for bitumen. IR and ^{13}C NMR spectra indicate that this sample has much greater aromatic–olefinic content than the other heavy oils. The IR spectrum has weak bands for olefins but a strong aromatic band at 1600 cm^{-1}. The integral for the NMR spectrum indicates 35% aromatic–olefinic content and 65%

saturate–alicyclic content. Molecular weights for this sample are lower than for the other heavy oils. All of the above data show that this heavy oil was very different from heavy oils produced later on during the experiment. Earlier discussion indicated that the light oil collected on the fourth day was also different from light oils collected later in the experiment. One possible explanation for differences of these two oils from their counterparts is that insufficient time from ignition has elapsed for representative materials to be produced.

Summary and Conclusions

Light oils produced during an in situ reverse combustion experiment near Vernal, Utah were upgraded considerably compared with the original bitumen. Some bases for this conclusion include greatly increased contents of components boiling in the gasoline, naphtha, and kerosine ranges, lower sulfur and nitrogen contents, decreased vanadium and nickel contents, and increased hydrogen-to-carbon atomic ratios.

Data related to heavy oils suggest that these products are very similar to the bitumen. This result is important because it shows that bitumen or very similar material can be produced during an in situ experiment. If the main product from recovery were bitumen and if the yield were high, most of the potential fossil energy value of a tar sand would be available above ground for additional processing.

The production of primarily bitumen-like heavy oil (55 barrels vs. 10 barrels of light oils) during this field experiment indicates that much of the tar sand within the pattern was not heated to high enough temperatures to cause major alterations in bitumen.

Acknowledgment

The aid of D. A. Netzel in obtaining and interpreting NMR spectra is acknowledged gratefully.

Literature Cited

1. Shea, G. D., Higgins, R. V., Wenger, W. J., Hubbard, R. L., Whisman, M. L., *U.S. Bur. Mines, Rep. Invest.* (1952) 4871.
2. Houghton, A. S., Howe, W. W., *Proc. World Pet. Congr., 7th, 1967,* 3, 703.
3. Gwynn, J. W., Ph.D. Dissertation, University of Utah (1970).
4. Wood, R. E., Ritzma, H. R., *Utah, Geol. Mineral. Surv., Spec. Stud.* (1972) 39.
5. Mauger, R. L., Kayser, R. B., Gwynn, J. W., *Utah, Geol. Mineral. Surv., Spec. Stud.* (1973) 41.
6. Bunger, J. W., "Shale Oil, Tar Sands, and Related Fuel Sources," ADV. CHEM. SER. (1976) 151, 121.
7. Bunger, J. W., Mori, S., Oblad, A. G., *Am. Chem. Soc., Div. Fuel Chem., Prepr.* (1976) 21(6), 1470.

8. Barbour, R. V., Dorrence, S. M., Vollmer, T. L., Harris, J. D., *Am. Chem. Soc., Div. Fuel Chem., Prepr.* (1976) **21**(6), 278.
9. Marchant, L. C., Land, C. S., Cupps, C. Q., *Energy Sources* (1975) **2**, 293.
10. Cupps, C. Q., Land, C. S., Marchant, L. C., *Am. Inst. Chem. Eng. Symp. Ser. 155* (1975) **72**, 61.
11. Land, C. S., Cupps, C. Q., Marchant, L. C., Carlson, F. M., *Petr. Soc. of CIM, Prepr.* (1976) Paper No. 7603.
12. Plancher, H., Green, E. L., Petersen, J. C., *Proc. Assoc. Asphalt Paving Technol.* (1976) **45**, 1.
13. Smith, N. A. C., Smith, H. M., Blade, O. C., Garton, E. L., *U.S. Bur. Mines, Bull.* (1951) 490.
14. Bean, R. M., Ph.D. Dissertation, University of Utah (1961).
15. Poulson, R. E., Jensen, H. B., Duvall, J. J., Harris, F. L., Morandi, J. R., *Anal. Instrum.* (1972) **10**, 193.

RECEIVED October 17, 1977. Mention of specific brand names or models of equipment is made for information only and does not imply endorsement by the Department of Energy.

Mass and Electron Paramagnetic Resonance Spectrometric Analyses of Selected Organic Components of Cretaceous Shales of Marine Origin

E. W. BAKER and S. E. PALMER—Departments of Chemistry and Biology, Florida Atlantic University, Boca Raton, FL 33431

W. Y. HUANG—University of New Orleans, Department of Biological Sciences, New Orleans, LA 70122

J. G. RANKIN—Shell Development Company, Bellaire Research Center, Houston, TX 77001

The organic maturity of Cretaceous Black Shales collected during the Deep Sea Drilling Projector (DSDP) drilling of the passive margins of the Atlantic has been assessed by combining two spectrometric techniques. Mass spectrometric analyses of porphyrins provide data on the following temperature-sensitive molecular transitions: aromatization, decarboxylation, carbocyclic ring rupture, transalkylation, and dealkylation. Insight into the origin and formation of petroporphyrins was provided by a special study of samples that had experienced an accelerated organic maturity caused by Miocene Cape Verde intrusive activity. Electron paramagnetic resonance (ERP) spectrometric analyses were used to determine the maturity of kerogens; EPR curves representing free radical formation pregeneration, generation, and post-generation are described.

The fossilization of organic matter involves five major steps: (1) microbial degradation, (2) condensation, (3) organic diagenesis, (4) thermal alteration (catagenesis), and (5) organic metamorphism (*1*). What methods are available to fit a given sample of unknown history into this sequence? A variety of empirical techniques have evolved from petroleum exploration which can be applied to this problem, but because most of these techniques have been developed specifically for source rock evaluation, the parameters generated are only applicable over a portion of the fossilization range. The application of combinations of spectroscopic techniques might be expected to bridge these limitations and allow more refined answers. We now present one such set, detail the assumptions, and show its applications in a particular geographical setting.

Kinetic theory tells us that the extent of completeness of a reaction is expressed by the integrated rate expression:

$$-\int_{C_1}^{C_2} dC = \int_{t_1}^{t_2} \int_{T_1}^{T_2} B\, e^{-\frac{A}{RT}}\, dt\, dT \qquad (1)$$

Where C = concentration of organic material (reactant)

   t = time

   T = temperature

   R = gas constant

   A = activation energy

   B = rate constant

The expression on the left of the equation represents simply the fraction of the reaction that is complete. This fraction then corresponds to the integrated time–temperature profile of the reaction. Thus, to obtain a particular fraction of completion, time can be traded for temperature or vice versa although temperature is obviously the more sensitive variable since it appears as an exponential term rather than a linear one. The activation energy determines the rate at any particular temperature. It must be recognized that an array of organic debris with various functionalities is present when considering geochemical reactions. Each functional group in a particular molecule then has a rate constant (B) and activation energy (A) (2, 3). Thus, to rigorously solve Equation 1 would require a quadruple integral including not only time and temperature but these additional variables. Today, the data are not available to even begin such a program and simplifying assumptions must be made.

An approach which yields useful data concerning the extent of completion of organic geochemical reactions is an empirical one. A class or group of compounds with constants and rates reasonably representative of the overall is chosen and its extent of reaction is determined over a particular geological period. The classes of compounds chosen were the tetrapyrrole pigments and kerogen: the first, represents primarily functional group changes in a single molecule and the second is representative of the reactions of a disordered polymer. Spectroscopic techniques applicable to the two classes are mass spectrometry (MS) for the porphyrins and electron paramagnetic resonance spectrometry (EPR) for the kerogen.

Spectrometric Methods

The spectra of the porphyrins were recorded on a DuPont 491–BR double focusing mass spectrometer using a solid probe inlet system. Details of the method have been described previously as well as the extraction and chromatographic purification of the pigment (4). In performing analyses of mixtures, observation of only the parent molecular ions without the complications of fragment ions is desired; therefore, low ionizing voltage (12–14 eV) is preferred. (Low voltage spectra give improved signal/noise ratios but at the expense of sensitivity).

Mass Distribution Parameters. The parent molecular ions of geoporphyrin mixtures fall into two major homologous series with peaks appearing in each series every 14 mass units. The distribution of molecular weights, termed the envelope, of the etio ($310 + 14n$, where n is an integer) and DPEP ($308 + 14m$, where m is an integer ≥ 2) series may be either a Gaussian distribution as in the case of petroporphyrins (up to 12 homologues) or a truncated series, characteristic of geoporphyrins in marine sediments (generally 4 major homologues). Thus the mass spectrum of a fossil porphyrin mixture may contain upwards of a score

Table I. Mass Spectrometric Parameters for Geoporphyrins

Parameter	Calculation
DPEP/etio[a]	Σ of I for each series
Weighted av. mass $(\overline{X})$[a]	$\Sigma IM/\Sigma I$
Band width (σ)[a]	$[\Sigma(IM^2)/\Sigma I - (\Sigma IM/\Sigma I)^2]^{1/2}$
Alkylation index (AI)[b]	$\Sigma[(\text{carbon number} - 32)I]/\Sigma I$
Mode[a]	I_{max}

[a] Ref. 17.
[b] Ref. 13.
[c] I = mass spectral peak intensity and M = mass.

of major peaks, possibly of several series. To be useful, the raw data must be condensed and simplified. To this end, a number of parameters have been defined. These parameters together with the method of calculation of each are listed in Table I. Simple computer programs are used to obtain the parameters after the peak intensities of each spectra have been normalized.

EPR Parameters. The EPR spectra of the kerogens were recorded on a Varian V-4500 spectrometer using the dual cavity technique (5). By this method highly accurate measurements of the usual EPR parameters (g, W, Ng) are possible since the nitroxide (Eastman 11293) and samples are scanned under identical conditions.

There are three EPR absorption lines in the standard nitroxide with spacing between adjacent peaks of 15.26 G and a g value of the middle line equal to 2.00603. This method is convenient in that the magnetic field can be calibrated accurately by the spacing between the adjacent peaks of the standard (15.26 G). In addition, the g value of the sample can be calculated simply by comparison of the sample to the standard according to the following equation: $H \times g = H' \times g'$, where $H = G$ reading of absorption line of same, $g = g$ value of sample, $H' = G$ reading of absorption line of standard, and $g' = g$ value of standard. The line widths of the EPR spectra were determined by measuring the distance between the two deflection points of the absorption line.

Kerogens were prepared in the usual manner by HCl and HF treatment of finely ground benzene-extracted sediments (6). No additional chemical treatment (e.g. $LiAlH_4$) was applied to remove pyrite. The samples were sealed in melting-point capillary tubes and heated in increments of time and temperature until readily observable changes in the spectral form, g value or band width occurred. A typical case for a mature kerogen required 250°C for 16 hr while heating at 150°C for 1 hr produced measurable changes for immature kerogens.

Relative Effects of Time and Temperature on Aromatization

Several factors which determine the state of diagenesis for a particular stratum are: geologic age, depth of burial, amount of thermal stress, and rate of subsidence. Naturally, the nature of the environment at the time of deposition also influences the kinds of degradation products found after time is allowed for diagenesis to occur. The level of biological productivity, redox potential of the sediments, and rate of burial of organic detritus are of particular importance. These factors determine the nature of the starting materials and the amount of organic matter trapped in the sediments.

Geochemically, chlorins are dehydrogenated quite readily to yield porphyrins (See Equation 2 for the most commonly referred to model structures and refer to Ring IV for the site of the hydrogen loss). Chemically, this transition is an aromatization and implies that rather deep-seated changes occurred in the molecule. Consequently this transition is viewed correctly as a significant turning point in the maturation process. Also in the geochemical transformation of chlorophyll to DPEP a number of reactions, some more and some less facile than dehydrogenation, occur; however, none of the other reactions are as readily followed analytically,

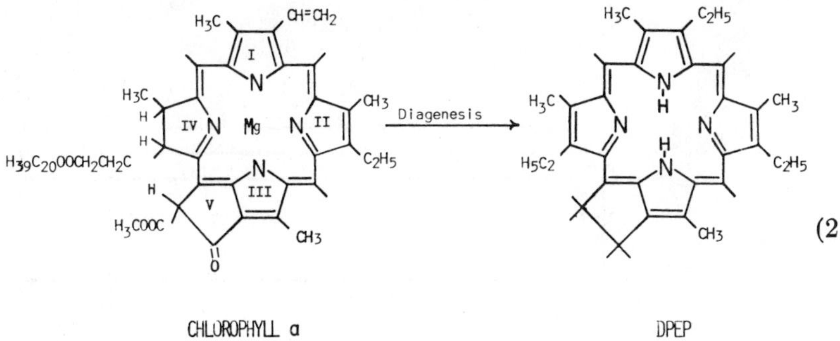

$$(2)$$

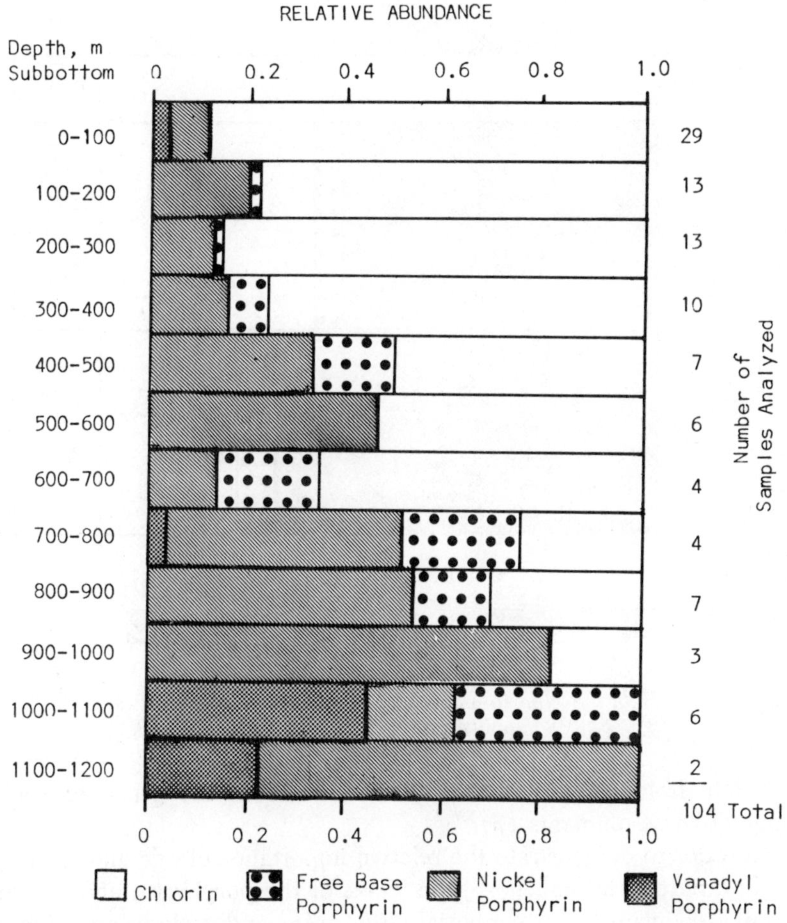

CHLOROPHYLL a   DPEP

Figure 1a. *Relative chlorin–porphyrin abundances in DSDP samples vs. depth of burial. Sample analysis and yields from Refs. 4, 6–16.*

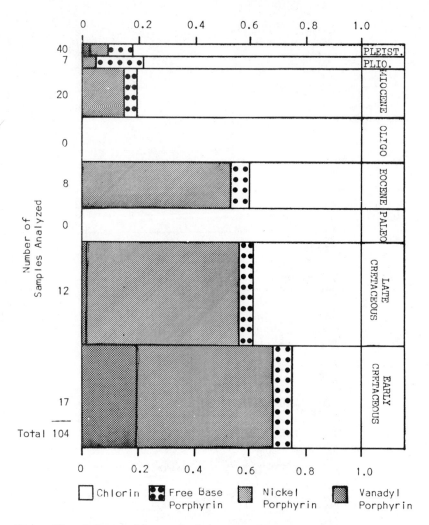

Figure 1b.   Relative chlorin–porphyrin abundances in DSDP samples vs. age. Sample analysis and yields from Refs. 4, 6–16.

since dehydrogenation (aromatization) results in the green–red color change of these pigments (7).

In order to demonstrate the relative importance of age and depth of burial on porphyrin diagenesis, the yields of the porphyrins and chlorins were plotted against depth of burial (Figure 1a) and geologic age (Figure 1b) (4, 6–16). In Figure 1a it is noted that as the depth of burial increases (from ocean floor to 1200 m subbottom) there is a systematic increase in the relative amount of the combined porphyrin components. Correspondingly, the chlorins decrease and finally disappear below the 1000-m sub-bottom. From Figure 1b, in which the relative yields of the green (chlorin) and red (porphyrin) pigments are plotted against geologic age of the sediment, a similar trend is apparent. With time the chlorins are converted slowly to porphyrins.

However, even in some of the oldest samples (Early Cretaceous) some chlorin is still present. Conversely, in the deepest samples no chlorin remains. Since increasing depth means increasing temperature (average gradient $=4°C/100$ m), cross comparison of these two plots show that temperature is more important than time in the chlorophyll diagenetic process, which is the general rule for organic maturation. The empirical conclusion is, of course, in accord with the theoretical ideas presented earlier concerning the factors which influence rates of reaction of buried organic matter.

Mass Spectrometric Indicators of Thermal Stress

Since temperature is more important than time, we now construct a model based on the changing molecular weight envelope of porphyrins relative to depth of burial/temperature (Figure 2). As a general rule, the complexity of the mass spectral envelope increases in direct proportion to the extent of geothermal stress that a particular stratum has undergone.

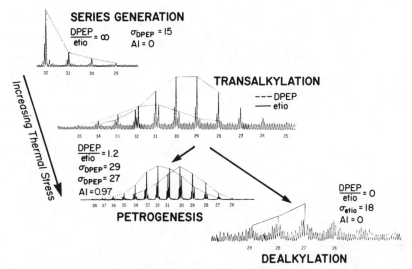

Figure 2. Schematic of tetrapyrrole pigment changes vs. age/temperature as discerned by mass spectrometry

During the conversion of chlorin to porphyrin, loss of methylene groups through decarboxylation and devinylation (and perhaps dealkylation) occurs and an asymmetric narrow envelope of the 308 $+14m$ series develops. The shape and width of this envelope is preserved after metallation with either nickel or vanadyl, remaining characteristic of relatively young metalloporphyrins. As thermal stress is applied, the envelope becomes wider and secondary peaks (the 310 + 14n series of etioporphyrins) begins to appear as shown in Figure 2. Development of the etio series is associated usually with nickel porphyrins; only with greater thermal stress are vanadyl etioporphyrins observed. As diagenesis gives way to catagenesis, the spectral envelope becomes very wide and the etioporphyrins develop into a distinct series. This is the case for porphyrins found in association with petroleum (petroporphyrins). In contrast, pigments that are exposed to severe temperatures (greater than 175°C) undergo dealkylation and are destroyed finally in the geological environment. Examples of mass spectra of a petroporphyrin and a dealkylated porphyrin are shown as contrasting end products at the bottom of Figure 2.

Changes within the mass spectral envelope can be described more clearly and simply in terms of the mass spectrometric parameters defined in Table I (17, 18). In response to increasing thermal stress, the parameters change in the following way: DPEP/etio ratio decreases, the weight average mass $(\overline{X})$ decreases, σ increases, and the alkylation index increases. Examples of the trends in these parameters as maturation proceeds in typical marine geoporphyrins and petroporphyrins are given in Table II.

A decrease in the DPEP/etioratio results from the opening of the carbocyclic (Ring V) (see Equation 3) and is detected mass spectrometrically by a shift from the 308 + 14m to the 310 + 14n series. (Three possible sites of ring rupture are shown in Equation 3.) The ratio trends from very large values to zero as diagenesis proceeds (Table II). The concerted loss of methylene units with opening of the carboxylic ring is

Table II.   Mass Spectrometric

State of Geoporphyrin Diagenesis	$\dfrac{DPEP\Sigma I}{etio\Sigma I}$
Immature free base or metalloporphyrin of DPEP series	∞
Metallated porphyrin under mild thermal stress	3.4
Dealkylated porphyrin	0
Transalkylated porphyrin	0.5
Petroporphyrin	1.2

[a] Data presented as the free base.

DPEP series
(308 + 14m)

etio series
(310 + 14n)

(3)

quantified by $\overline{X}$. Because mass loss and ring opening are related, $\overline{X}$ of the etio series is consistently lower than $\overline{X}_{DPED}$ by one to two methylene units (14 to 28 mass units) as shown in Table II. Other mechanisms by which side chain carbons are lost (or gained) are also operative giving rise to porphyrin of C_{28} and lower carbon number (*see* mass spectrum of such a case in Figure 2 labeled dealkylation). The example given is a nickel porphyrin fraction in which only a few members of the etio series remain. Thus, the DPEP/etio ratio is zero and the band width is narrow ($\sigma = 18$). In time, these side chain alteration mechanisms (either gain or loss) give rise to a wider range between the highest and lowest mass porphyrin; thus, σ, the half band width of the best fit Gaussian curve, increases. Substantial differences in band width are seen between values of immature, mildly stressed porphyrins of the DPEP series ($\sigma = 15$–18) and those that are transalkylated ($\sigma = 27$). The alkylation index (A.I.) was designed to indicate the dergee of transalkylation that has occurred in a given metalloporphyrin mixture (Table I) (*13*). Normally, porphyrins found in marine sediments are nonalkylated and have a value of zero; however, A.I. values of 0.08–0.7 have been observed for thermally stressed porphyrins (Table II). For comparison, a highly alkylated petroporphyrin of the DPEP series such as Boscan (Cretaceous) with six series members above C_{32} has an alkylation index of 0.97 (*17*). The DPEP

Parameters of Selected Geoporphyrins[a]

Weighted Average Mass		Band Width		Alkylation Index[b]
$\overline{X}DPEP$	$\overline{X}etio$	$\sigma DPEP$	$\sigma etio$	
466		15		0
454	422	18	26	0
	420		18	
477	458	27	30	0.70
482	469	27	29	0.97

[b] Based on peak intensities of the DPEP series.

series is chosen for computation of the index because the etio series generally is biased towards the lower molecular weight members.

EPR Indicators of Maturity

A second model, based on the EPR of kerogens, was selected to provide a framework into which the measures of maturity of high molecular weight compounds could be fitted. This choice was based initially on the well known fact that most kerogens (1) give a very strong EPR signal generally with a single absorption in the range of $g = 2.0033 \pm .00100$ and a band width of 8 ± 3 G and (2) that the parameters (g, W, and Ng) react at least in a limited way to structural transformations (19, 20, 21). As experimental data accumulated several facts emerged. First, the utility of the usual parameters is limited even in those straight-forward cases where a single absorption line is observed. For example, in some suites of samples, variations (trends?) in g value occur only in the fourth decimal place even though the geological change is substantial (22), and accurate measurement of Ng is so tedious and time consuming, that it is precluded on any but a limited number of key samples. Secondly, the most interesting cases turn out to be those where multiple unresolved absorption lines are present. Figure 3 presents three cases of atypical spectra in which such multiple unresolved absorption lines are present. In these and similar cases the parameters (g, W, and Ng) cannot be obtained without using a sophisticated deconvolution technique. For these reasons, we have chosen to depend mainly on the visual form of the EPR spectra. Some of the forms that we have recognized as standing at key points in the reaction series are given schematically in Figure 4. Band width (but not intensities) of all the spectra in Figure 4 with the exception of "active" kerogen can be compared since all were recorded at essentially identical instrument settings with the nitroxide standard (spacing between adjacent peaks = 15.26 G) juxtaposed only for asphaltene. On the other hand, the active kerogen spectrum was recorded at a span five times as large (again the nitroxide spectrum is juxtaposed) because of the unusually wide band width.

Figure 4 shows that at the generation stage the sequence branches into a high hydrogen-to-carbon ratio product (petroleum) and a low hydrogen to carbon ratio product (kerobitumen). Since some forms in this sequence are identical in appearance, the location in the sequence can be distinguished only by the transitions occurring when the samples are subjected to laboratory heat treating experiments. Note that typical pregeneration kerogen and post-generation kerobitumen (23) exhibit nearly identical EPR spectra (see Figure 4). Upon heating, line narrowing would indicate that the sample was post-generation contrasting with the

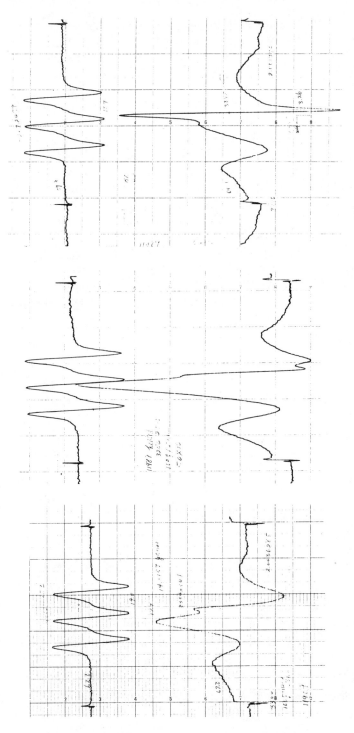

Figure 3. EPR spectra of three selected kerogens showing complex (atypical) spectra. Above each sample spectrum is a reference spectrum of the nitroxide standard (4-acetamido-2,2,6,6,-tetramethyl piperidine-1-oxyl). The g value of the middle line is 2.0061 and splitting is 15.26 G.

line broadening which would indicate a pregeneration state being pushed towards the "active" kerogen of generation. The second post-generation product, asphaltene, which is a fraction of petroleum, gives a characteristic EPR spectrum shown as the left branch in Figure 4.

In some selected cases the g value and peak width can be related to maturity/thermal stress. One such case, where a column of generally homogeneous organic material was present and gave smoothly trending parameters, is shown in Table III.

Incipient radical formation (upper two spectra shown in Figure 4) was discovered in sediment kerogen from DSDP Site 367 in the Cape Verde Basin (5). DSDP Site 368 found a basaltic sill intruding into a Cretaceous shale and thus provided severely thermally treated coke shown as the "end of the line" in Figure 4. The model for the "active" kerogen (i.e. that which is in the petroleum generation state) is a sample of a known petroleum source rock kerogen. Representative forms are shown for the pregeneration, asphaltene, and kerobitumen cases.

The use of "visual forms" restricts the method to qualitative aspects. Furthermore, it is cumbersome to store, retrieve, and transmit the data.

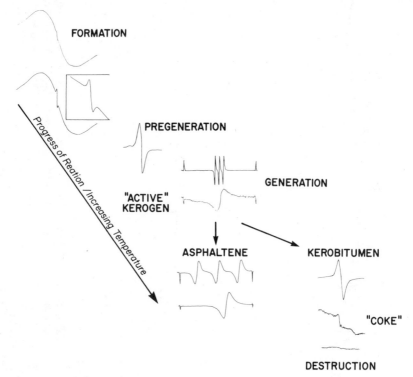

Figure 4.   Schematic of EPR spectra vs. progress of maturation. See text for details.

Table III. EPR Parameters of Kerogens Isolated from Core
Samples in Proximity of Dibase Sill, Site 368, Leg 41

Distance from Sill(m)	$(g - 2) \times 10^4$	Peak Width (G)
+10.00	37.0	10.23
+ 7.37	32.9	9.46
+ 4.80	29.4	6.46
+ 3.60	30.1	3.25
+ 2.43	33.5	19.32
+ 1.88		
+ 1.33		
+ .65		no signal
+ .05		
− .17		
− .71		
− 1.21	30.1	5.52
− 1.93	30.4	6.69
− 2.67	31.2	6.40
− 3.67	31.4	7.34
− 4.93	32.5	8.36
− 7.41	31.2	6.12
− 8.39	36.3	10.28

Extension and expansion of the method will require that suitable parameters be defined which describe and quantify the information contained in the shapes of the absorption lines just as special mass spectrometric parameters (Table I) were required for geoporphyrins.

Response of Parameters to a Steep Geothermal Gradient

Although reaction rates can be measured in the laboratory with prediction of products generated from known reactants under controlled time–temperature conditions, few test cases have been naturally available. We now describe such a case in which, because of an igneous intrusion into an organic-rich marine shale, the response of geoporphyrins to a steep thermal gradient was measured (*13*). The temperature being measured is that which left a trace of its presence observable as changes in the organic matter of the shale.

Mass spectrometric parameters of a typical immature metalloporphyrin fraction are given in Table II. These parameters were used to interpret mass spectra of porphyrin samples recovered within known distances from a 13.5 m diabase sill (Table IV). The complexity imposed by temperature-controlled reactions (transalkylation and conversion of DPEP to etio series) can cloud trends within an individual parameter, but used together, the parameters permit definition of these reactions. As the sill is approached, the DPEP/etio ratio approaches zero for both

Table IV.   Molecular Weight Distribution Parameters of
of the Dibase Sill,

Distance Above (+) and Below (−) Sill (m)	Porphyrin Type	$DPEP\Sigma I$ $etio\Sigma I$
+15.87	nickel	7.5
+7.37	nickel nd^b	
+4.80 – +0.65	no pigment	
−0.17 – −1.93	no pigment	
−2.67	nickel nd vanadyl nd	
−3.67	nickel vanadyl nd	0
−4.69	nickel vanadyl	0.5 0.5
−4.93	nickel vanadyl	1.5 1.1
−7.41	nickel vanadyl	3.5 7.9
−8.39	nickel	2.1

[a] Data reported as the free-base porphyrin.

nickel and vanadyl porphyrins; this trend is interpreted as rupture of the carbocyclic ring. Weighted average mass tends toward lower values indicating an average loss of one or two methylene groups. In contrast, both the band width and alkylation index go through a maximum at 4.93 m from the sill, perhaps representing best conditions for transalkylation. As the distance from the sill decreases from 4.93 to 3.67 m, dealkylation reactions outweigh transalkylation. This is particularly noticeable at 3.67 m where only an extremely dealkylated nickel etioporphyrin series series was present. No pigment survived close to the sill (0.17–1.92 m).

These data expose some pertinent new facts concerning porphyrin diagenesis as well as shedding light on the origin of petroporphyrins which have long been believed to be the products of chlorophyll diagenesis (24, 25). Temperature is the promoter of both transalkylation and dealkylation reactions which can occur in deeply buried sediments. However, because of the extreme temperatures (300°–1000°C) to which the sediments in this case were exposed, dealkylation was favored over transalkylation. Under milder conditions (50°–150°C) the transalkylation

Metalloporphyrins Isolated from Core Samples in Proximity Site 368, Leg 41[a]

Mode		Weighted Average Mass		Band Width		Alkylation Index
DPEP	etio	$\overline{X}$DPEP	$\overline{X}$etio	σDPEP	σetio	
462	436	449	441	18	21	0
	408		420		18	0
462	436	462	436	18	20	0.1
448	436	463	441	25	24	0.5
462	436	463	446	23	24	0.3
476	464	476	456	27	30	0.7
462	450	465	453	22	23	0.3
476	478	467	459	16	20	0.1
476	450	464	439	18	27	0.08

[b] nd = No mass spectrometric data.

reaction might have prevailed leading to typical petroporphyrins having parameters similar to the example given in Table II.

Relative stabilities and reactivities of the two metalloporphyrins can be observed by comparison of the parameters. The alkylation index shows that vanadyl chelation promotes transalkylation relative to nickel. On the other hand, nickel porphyrins are degraded more rapidly through the loss of methylene groups and/or etio series generation than vanadyl porphyrins. These differences between the two chelates are best observed by comparison of the series mode and weighted average mass shown in Table IV.

The distribution of vanadyl vs. nickel porphyrins is noteworthy (Table IV). An overview of the data from Site 368 shows that normal porphyrin diagenesis has progressed to the nickel DPEP porphyrin stage at + 15.87 m (*13*). At greater depth, the normal process is interrupted by high thermal stress with concomitant formation of vanadyl porphyrins and decomposition of the native nickel porphyrins. Further below the sill (−8.39 m) only slightly altered nickel porphyrins, again predominantly of the DPEP type, are observed. The occurrence of vanadyl porphyrins in only highly stressed sediment samples may suggest that their formation is related to the thermal input provided by igneous intrusion. Whether vanadyl porphyrins were generated in situ, i.e. being released from kerogen, is presently an open question. However, since migration or reworking do not fit the data, in situ generation seems to be the best explanation.

The trends occurring in two EPR parameters (g and W) over the range of 10 m above to 8 m below the sill are given in Table IV (5). Below the sill the trend is clear, with band width and g value decreasing with increasing thermal stress, though not linearly nor without significant deviation. Possibly the sample at − 7.41 m is indicative of the kind of counter trend which results from nonhomogeneity of the organic material in the sediment column. In fact, such nonhomogeneity is probably typical rather than atypical and so precludes any absolute relationship between such parameters and temperature as has been suggested (19). Above the sill the variations are erratic because of a number of minor intrusions which produced an incomprehensible thermal profile.

Intrusions, Diagenesis, and Temperatures

Using the data from Table IV, conventional knowledge of reaction kinetics, principles of igneous geology, and intuition, a temperature profile in the region of the sill has been constructed and is shown in Figure 5. In constructing the profile, it was necessary to make a number of assumptions of varying quality which in turn reflect as ranges of uncertainty in the most likely profile. The most likely profile is shown as a curve with the range of uncertainty superimposed as a shaded band. Since it was obvious from the porphyrin data that the temperature increased sharply as the sill was approached, a logarithmic dependence of temperature vs. distance was assumed intuitively. Sample characteristics (aromatics present, pigment present, degree of conversion to etioporphyrins, and transalkylation) were noted and are listed as a guide in Figure 5. Temperature ranges for each sample site below the sill were estimated then based on the half-life for pigment destruction and for DPEP to etio conversion.

Samples taken within 2 m of the sill varied from light brown at − 0.17 m to black at − 1.93 m. The black samples contained some extractable organic material and coke; however, essentially all of the vitrinite was coked and no organic material was detected in the light brown sample. On this basis temperature limits of 880° and 1060°K are estimated (26). The upper limit of the magma was set at the usual geology textbook temperature of 1473°K (1200°C). The estimate of the

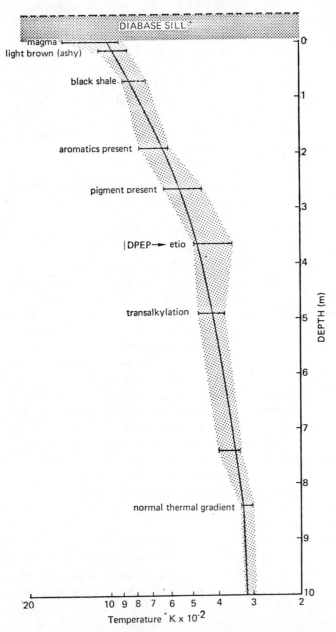

Figure 5. Proposed reconstruction of thermal gradient in the immediate vicinity of 13.5-m Miocene diabase sill in Cretaceous black shale. Assumptions made at incremental distances are detailed in text.

temperature at the point of contact of the sediment and magma was obtained by averaging, i.e. $(1473°K + 310°K)/2 = 890°K$ or $617°C$. However, this estimate may be high since contact metamorphism at 1000-m depth of burial would occur at $300°–800°C$ (27) and the alteration zone was about 1 –1.5-m thick.

Below this baked zone, the characteristics of the organic extracts from each sediment sample could be used to estimate the temperature. For example, at $- 1.93$ m, the maximum proportion of aromatic compounds were found but no pigment was present. The first observation defines the upper temperature limit at about $773°K$ and the second—the minimum—at $600°K$.

A sample at $- 2.67$ m contained a trace of extractable pigment (0.2 and $0.4 \mu g/g$ of nickel and vanadyl porphyrin, respectively), while a sample some 60 cm closer to the sill was devoid of pigment presumably because it was overheated and destroyed. Conversely, the sediments slightly further from the sill contained significantly greater amounts of extractable pigment (see Table IV). Based on pigment yield and the data of Rosscup and Bowman (28) the maximum momentary temperature limit for this location is placed at $350°C$ in the following way. According to the data of Rosscup and Bowman, at $373°C$ the half-life of nickel petroporphyrin is 15 hr and that of vanadyl petroporphyrin 8.5 hr. Assuming, as is usually done, that ten half-lives will reduce the concentration below detectable limits, at $350°C$ the porphyrin at this location would have been destroyed in a matter of days and certainly temperatures near the maximum persisted longer than days near a 13.5-m sill.

To estimate the lowest temperature maximum which might have been reached soon after the intrusion occurred, some insight into the time–temperature profile is necessary and some assumptions are required concerning the rate at which crystallization and cooling of the sill took place. Daly (29) gives estimates for the crystallization of magma in contact with air of 3 and 300 yr for sill thicknesses of 10 and 100 m. Direct extension of these estimates to a subterranean sill is unwarranted but if applied cautiously they can provide the order of magnitude.

Making the standard assumption that the reaction rate decreases by a factor of 2 for each $10°C$ reduction in temperature, straightforward calculations show that 50,000 yr at $170°C$ would eliminate the pigment from the record (e.g. reduce the concentration to a nondetectable level).

To reiterate, if one assumes a temperature of $350°C$, then the time at that temperature could have not persisted longer than 60–600 hr (0.05 yr) or else the traces of pigment ($0.2 \mu g/g$) would not be present. Conversely, if one assumes a temperature of $170°C$, in somewhat less than 50,000 yr the pigment concentration would be reduced to the observed level. Extension of Daly's data suggests neither a time as short as 0.05

year or as long as 5×10^4 yr but rather gives a best estimate of 5 yr for the cooling of a 13.5-m sill. This best estimate of the time in turn gives a best estimate of the temperature ($550°K$ ($282°C$)) with the range of $446°K$ ($170°C$)–$623°K$ ($350°C$) as shown by the bars in Figure 5 and the labeled pigments present.

The maximum transient temperature at the points further from the sill are determinable by yet another characteristic of the porphyrins, namely the opening of the isocyclic ring of DPEP to form porphyrins of the etio series (Equation 3). Didyk et al. (*30*) performed experiments which showed that the DPEP to etio conversion is approximately 50% completed after 150 hr at $483°K$ ($210°C$). Making the same assumptions as previously, (1) that the rate decreases by a factor of 2 for each $10°C$ reduction and (2) that there is a reaction time of 4×10^4 hr (5 yr) (based on an observed DPEP to etio ratio of 0.5) the best estimate of the temperature at -4.69 m is $414°K$ ($140°C$). Furthermore, to be consistent with these data, the temperature at -4.69 m could not have exceeded $483°K$ for as long as two half-lives (300 hr). Assuming total conversion to etio (destruction is slow at this temperature), one half-life by definition gives a DPEP/etio of 1.0 and two half-lives, a DPEP/etio ratio of 0.33 bracketing the observed ratio of 0.5. Likewise, making the assumption that at -3.67 m the concentration of the DPEP series has been decreased to just below the detectable level, the maximum, minimum, and most probable temperatures are $485°$, $350°$, and $472°K$.

Mass spectrometric analyses (Table III) show that at -4.93 m, the DPEP/etio ratio of the nickel porphyrin fraction was 1.5 while at -7.4 m the ratio was 3.5 and for the vanadyl porphyrin fraction ratios of 1.1 and 7.9 were found at the same distances. Using the same assumptions and calculations as above, a most probable temperature at each distance can be estimated, which combined with time ranges, are shown by the uncertainty bars in Figure 5 using only these two points. An intuitive confirmation of the calculations may be made as follows. The observed difference between the vanadyl porphyrin DPEP/etio ratio of 1.1 at -4.93 m and 7.9 at 7.4 m corresponds roughly to 3 doublings or a $30°C$ temperature difference if it is assumed that a $10°C$ reduction lowers the conversion by 50%. This argument disagrees slightly with the overall best fit line in which the temperature difference is $50°$–$60°C$.

As mentioned earlier, transalkylation, measured both by band width and A.I., was most extensive in the sample from -4.93 m (Table IV). No kinetic data for the transalkylation reaction are available; however, intuitively a maximum in the range of $100°$–$150°C$ seems reasonable. Below that range the alkylation reaction is noncompetitive with other diagenetic reactions and above that range dealkylation and destruction predominate.

A sample taken at −8.4 m contained no vanadyl porphyrin and contained a nickel porphyrin fraction with an A.I. of 0.08. These observations represent the maturation expected for a Cretaceous sediment at a 100-m depth of burial. A gradient of 4°/100 m gives a probable temperature of 313°K (40°C) in the absence of the sill. The above indices of maturation indicate that an excursion from this temperature at this point was limited and transient. Customarily, igneous bodies affect organic material in intruded sediments to a distance about twice their thickness which would be about 27 m. In this case, it would appear that significant changes in the organic materials were restricted to a region considerably narrower.

Conclusions

The combination of two techniques that measure independent components of the same geological environment permits more dependable interpretations than when used separately. The changes in mass spectrometric parameters, particularly an increase in band width and A.I. and a decrease in DPEP/etio ratio, as well as changes in the EPR signal can be used collectively to discern onset through completion of thermally dependent reactions. While the mass spectral method gives a detailed resolution of many of the specific reactions involved, EPR provides a better overall view of maturation and covers some areas inaccessible to the former.

The studies reported here provide information pertinent to current discussions of tetrapyrrole diagenesis. Systematic analyses of marine sediments of known geologic age and depth of burial have shown that chlorins give way to free-base porphyrins with aromatiaztion complete at approximately 1000-m depth of burial (Figures 1a and 1b). Initial homologous series generation of up to four members (C_{32}–C_{29}) occurs during the chlorin–porphyrin transition (Figure 2); proposed mechanisms involve decarboxylation and devinylation. Metallation with either nickel or vanadyl is possible; however, as shown in Figures 1a and 1b nickel porphyrins have a wider distribution.

As diagenesis progresses, rupture of the isocyclic ring of porphyrins of the DPEP series results in generation of the etio series. The rate of conversion is faster in porphyrins chelated with nickel than with vanadyl. Transalkylation of both metalloporphyrin species have been observed; however, vanadyl promotes this reaction relative to nickel. Cleavage of alkyl groups (dealkylation) occurs at elevated temperatures (177°–312°C) which are not favorable for transalkylation. Observation of dealkylation can be only transitory since the process leads ultimately to destruction of the pigment (Figure 2).

In conclusion, the geoporphyrins are the only likely precursors to the petroporphyrins. Proposed mechanisms by which specific chlorophylls, the biological precursors of geoporphyrins, can give rise to high molecular weight members should be considered as hypothetical. Petroporphyrins contain no information that relates a particular molecular weight species to a particular chlorophyll.

We conclude that it is possible to determine the maturity of a sediment by examination of the properties of kerogen and tetrapyrrole pigments. In fact, a fairly detailed reconstruction of the chemical events surrounding a Miocene intrusion into a Cretaceous shale has been possible.

Acknowledgments

This research was supported by the Oceanography Section of the National Science Foundation, Grants GA-37962, GA-43359X, DES 74-12438 AO1, and OCE 74-12438 AO2.

Literature Cited

1. Hunt, J., "Advances in Organic Geochemistry 1973," B. Tissot, F. Bienner, Eds., pp. 593–605, Editions Technip, Paris, 1974.
2. Connan, J., *Am. Assoc. Pet. Geol. Bull.* (1974) **58**(12), 2516.
3. Waples, D., Connan, J., *Am. Assoc. Pet. Geol. Bull.* (1976) **60**(5), 884.
4. Baker, E. W., "Initial Reports of the Deep Sea Drilling Project," R. G. Bader, et al., Eds., Vol. 4, pp. 431–438, U.S. Government Printing Office, Washington, 1970.
5. Baker, E. W., Huang, W. Y., Rankin, J. G., Castaño, J. R., Guinn, J. R., Fuex, A. N., "Initial Reports of the Deep Sea Drilling Project," Y. Lancelot, E. Siebold, et al., Eds., Vol. 41, pp. 839–849, U.S. Government Printing Office, Washington, 1978.
6. Robinson, W. E., "Organic Geochemistry," Eglinton, G., Murphy, M. T. J., Eds., pp. 181–195, Springer–Verlag, New York, 1969.
7. Baker, E. W., "Organic Geochemistry," G. Eglinton, M. T. J. Murphy, Eds., pp. 464–497, Springer–Verlag, New York, 1969.
8. Baker, E. W., Smith, G. D., "Initial Reports of the Deep Sea Drilling Project," Heezen, B. C., MacGregor, I. D., et al., Eds., Vol. 20, pp. 943–946, U.S. Government Printing Office, Washington, 1973.
9. Baker, E. W., Smith, G. D., "Initial Reports of the Deep Sea Drilling Project," D. E. Karig, J. C. Ingle, et al., Eds., Vol. 31, pp. 905–909, U.S. Government Printing Office, Washington, 1975.
10. Baker, E. W., Smith, G. D., "Initial Reports of the Deep Sea Drilling Project," D. E. Karig, J. C. Ingle, et al., Eds., Vol. 31, pp. 629–632, U.S. Government Printing Office, Washington, 1975.
11. Baker, E. W., Palmer, S. E., Parrish, K. L., "Initial Reports of the Deep Sea Drilling Project," M. Talawani, G. Udinstev, et al., Eds., Vol. 38, pp. 785–789, U.S. Government Printing Office, Washington, 1976.
12. Baker, E. W., Palmer, S. E., Huang, W. Y., "Initial Reports of the Deep Sea Drilling Project," H. M. Bolli, W. B. F. Ryan, et al., Eds., Vol. 40, U.S. Government Printing Office, Washington, in press.
13. Baker, E. W., Palmer, S. E., Huang, W. Y., "Initial Reports of the Deep Sea Drilling Project," Y. Lancelot, E. Siebold, et al., Eds., Vol. 41, pp. 825–837, U.S. Government Printing Office, Washington, 1977.

14. Baker, E. W., Palmer, S. E., Huang, W. Y., "Initial Reports of the Deep Sea Drilling Project," D. Ross, Y. Neprochnov, et al., Eds., Vol. 42B, U.S. Government Printing Office, Washington, in press.
15. Palmer, S. E., Huang, W. Y., Baker, E. W., "Initial Reports of the Deep Sea Drilling Project," B. Tucholke, P. Vogt, et al., Eds., Vol. 43, U.S. Government Printing Office, Washington, in press.
16. Baker, E. W., Palmer, S. E., Huang, W. Y., "Initial Reports of the Deep Sea Drilling Project," W. E. Benson, R. E. Sheridan, et al., Eds., Vol. 44, U.S. Government Printing Office, Washington, in press.
17. Baker, E. W., Yen, T. F., Dickie, J. P., Rhodes, R. E., Clark, L. F., *J. Am. Chem. Soc.* (1967) **89**, 3631.
18. Baker, E. W., Palmer, S. E., "The Porphyrins," D. Dolphin, Ed., Vol. 1, pp. 485–551, Academic, New York, 1978.
19. Hwang, P. T. R., Pusey, W. C., United States Patent **3,740,641** (1973).
20. Ishiwatari, R., Ishiwatari, M., Rohrback, B. G., Kaplan, I. R., *Geochim. Cosmochim. Acta* (1977) **41**(6), 815.
21. Yen, T. F., Sprang, S. R., *Geochim. Cosmochim. Acta* (1977) **41**(8), 1007.
22. Baker, E. W., *Conference on Fossil Fuel Chemistry and Energy, Laramie, Wyoming, July 22–26, 1975.*
23. Erdman, J. G., "Petroleum and Global Tectonics," A. G. Fischer, S. Judson, Eds., p. 225–248, Princeton University, Princeton, NJ 1975.
24. Treibs, A., *Angew. Chem.* (1936) **49**, 682.
25. Corwin, A. H., *World Pet. Cong., Proc., 5th, New York* (1959) Paper **V-10**.
26. Bostick, N. H., Ph.D. Dissertation, Standford University (1970).
27. Rolfe, W. D. I., Brett, D. W., "Organic Geochemistry," G. Eglinton, M. T. J. Murphy, Eds., pp. 213–244, Springer–Verlag, New York, 1969.
28. Rosscup, R. J., Bowman, D. H., *Am. Chem. Soc., Div. Pet. Chem., Prepr.* (1967) **12**, 77.
29. Daly, R. M., "Igneous Rocks and the Depths of the Earth," p. 63, McGraw–Hill, Highstown, NJ 1933.
30. Didyk, B. M., Alturki, Y. I. A., Pillinger, C. T., Eglinton, G., *Nature (London)* (1975) **256**, 563.

RECEIVED August 5, 1977.

A Preliminary Electron Microprobe Study of Green River and Devonian Oil Shales

E. A. HAKKILA, N. E. ELLIOTT, J. M. WILLIAMS,
and E. M. WEWERKA

Los Alamos Scientific Laboratory, Los Alamos, NM 87545

One sample each of Green River shale from Colorado and Devonian shale from West Virginia was examined with the electron microprobe. These studies demonstrated that the identities and distributions of many shale minerals can be inferred by correlating elemental associations with x-ray mineralogical data. Dark bands in both samples were composed primarily of organic matter associated with quartz, feldspars, clays, and carbonates. The carbonates present in the dark band were identified as calcite and dolomite which contained small amounts of iron. Dolomite, iron-rich dolomites and magnesites, and small amounts of siderite predominated in the light organic-poor bands in these specimens. Organically associated sulfur was observed only in the Green River shale sample.

The energy crisis, brought about by dwindling oil and gas reserves, has prompted our nation to seek new sources of these strategic commodities. Among the possible oil and gas sources are the Green River shales of the Rocky Mountains and the Devonian shales in the eastern United States. The organic matter (kerogen) in the Green River shale is the potential source of over 4 trillion barrels of oil or petroleum feed stock (1). In addition to the kerogen entrapped within the Devonian shale there is a huge resource (over 500 quadrillion cu ft) of natural gas (2). Unfortunately, the resources in both these deposits are contained in rock matrices, and extraction of the oil or gas from either of these shale beds is costly and very difficult.

While some efforts are underway to increase production efficiency, others have been directed at characterizing the rock matrices to determine how the resources are contained. This knowledge will be useful

0-8412-0395-4/78/3-170-181$05.00/1

in deciding how the oil or gas can be produced. The purpose of this study is to demonstrate the utility of the electron microprobe in characterizing energy bearing shales. For this purpose, we have chosen a sample from each of the major U.S. shale areas. This initial study, although involving only two samples, still allows some comments to be made about the compositional similarities and differences between the two shale types.

Instrumentation and Procedure

One sample each of Green River and Devonian shales was obtained from the Laramie and Morgantown Energy Research Centers, respectively. The Green River shale was taken from the Mahogany Zone in the Piceance Basin of Colorado; the Devonian specimen was taken from the Marcellus layer in Jackson County, West Virginia. Each sample was sectioned as a slab, 2.5–3.0-cm long and 1.2-cm wide by 0.6-cm thick, with the long direction parallel to the vertical axis of the core. The Devonian sample was pressure-impregnated with epoxy resin to preserve its structure during sample polishing. The samples were polished to reveal the microstructure and then coated with a 100-Å layer of carbon to provide electrical conductivity.

The x-ray analyses were obtained using powder camera and diffractometer techniques. In the former, powdered shale was loaded into a glass capillary which then was exposed to 20 hr of $Cu(K\alpha)$ radiation. The x-ray pattern was taken with a standard Norelco 114.6-mm diameter powder camera equipped with a nickel filter.

For the diffraction technique, −325-mesh shale powders were packed into the 7/8-in. diameter by 1/16-in. deep cavity of a 1-in. diameter aluminum holder and pressed smooth with a glass plate. A Norelco–Phillips diffractometer equipped with a $Cu(K\alpha)$ source was used to analyze the samples. The diffractometer was driven at 1°/min and the intensity response of the nonrotating sample recorded on a strip chart recorder. Peak assignments in both cases were made by visual comparisons with standards.

The electron microprobe used in this study was an Applied Research Laboratories Model EMX–SM, equipped with three wavelength dispersive x-ray spectrometers of differing wavelength capabilities and with a silicon(lithium) $(Si(Li))$ energy dispersive spectrometer. This arrangement allowed all elements heavier than beryllium (Be) to be studied. The spectrometers, stage motion (x- or y-direction), and scaler–timer readout were controlled manually or through a PDP–8E computer. The energy dispersive spectrometer was used for rapid identification of elements in areas or grains; however, the detailed studies were performed with the wavelength dispersive spectrometers which give higher sensitivity. A secondary electron detector permitted scanning electron photomicrographs (SEM) to be taken. These SEM photographs allowed the two-dimensional x-ray images of elemental distributions to be correlated with surface features.

Two techniques were used to obtain information on sample composition. In the first, the electron beam was rastered rapidly over a specified area (here 80 × 100 μm) and the characteristic x-rays of the elements sought were monitored. An x-ray intensity ratio for each element in the entire area was determined from the intensity of the measured x-rays vs. those from a standard material: pure iron for iron, pyrite for sulfur, pure aluminum for aluminum, a grossularite for calcium, and graphite for carbon. By sweeping many adjacent areas, detailed information about the microscopic changes in elemental and mineral composition along the total length of a sample could be obtained. In the second technique, the electron beam was rastered slowly across a selected area of the sample, and the intensities of the elements of interest were recorded photographically from an oscilloscope screen. This method provided point-by-point information about the elements present and rough estimates of their relative concentrations and distributions. Elemental data from the microprobe analyses were correlated with the x-ray mineralogical analyses to provide the basis for inferring the micromineralogy of these two samples.

Results and Discussion

The Green River oil shale is an organic-rich marlstone that is thought to have been deposited in a lacustrine environment (3, 4). These shales are composed of distinct pairs of light and dark bands, called varves, that reflect seasonal variations in the organic content of the shale (5). The varves in the Green River shale were shown by Smith to average about 20 μm in thickness (3).

Dolomite, quartz, and analcite were the major minerals found in our shale sample. Lesser amounts of calcite, albite, K-feldspars, K-clays, and mica-clays were observed as well as trace levels of siderite and either pyrite or marcasite. This mineralological composition is similar to that reported elsewhere for the Green River shales (3, 4).

The banded structure of the Green River shale studied here is shown in the photomicrograph of Figure 1. Using the rapid rastering microprobe technique, the entire 3-cm length of the sample was analyzed at 100-μm intervals for Fe, S, Ca, Al, and C. The ratios of the measured x-ray intensities to those for the standards for these elements are plotted in Figure 1. The dark shale areas are enriched considerably in organic matter as indicated by the higher intensity C x-rays. The elemental analyses indicate that the light bands, on the other hand, are composed

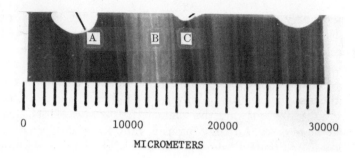

MICROMETERS

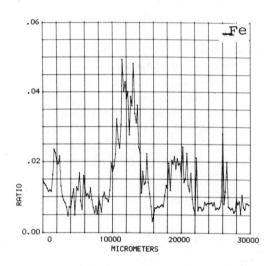

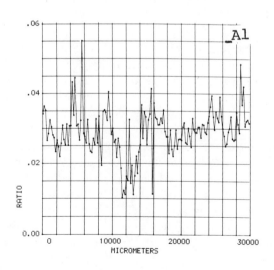

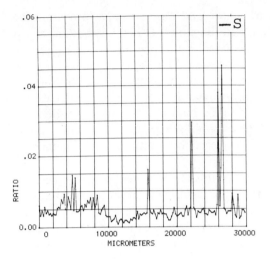

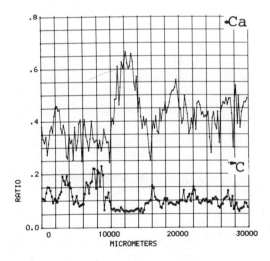

Figure 1. Distributions of Fe, Al, S, Ca, and C along length of Mahogany zone shale. Letters on photomicrograph refer to areas described in text.

primarily of carbonates, and they are seen to contain less S and Al and more Ca and Fe than the darker C-rich bands. The sharp spikes in the S intensities are associated with areas of high Fe (presumably pyrite or marcasite) concentrations. These pyritic areas are associated mainly with the organic-rich bands. A notable exception to the Fe/S association is the S spike at approximately 16,000 μm which is not accompanied by a corresponding rise in Fe intensity and which indicates the presence of nonpyritic S. These observations are consistent with patterns of geochemical genesis proposed by Smith for the formation of Green River shale (3).

From the areas of this Green River shale sample that were examined in detail using the slow rastering microprobe technique, we have chosen three for discussion. Area A was a dark band located near the 6000-μm position in Figure 1. Area B was in the wide, lighter colored area located at $\sim$ 13,000 μm, and area C included a black, organic strip at $\sim$ 16,000 μm.

A scanning electron photomicrograph (SEM) and corresponding element x-ray intensity for area A are shown in Figure 2. This area contains numerous particles, which appear as light, often angular, crystals in the SEM. These particles contain only S and Fe in detectable concentrations, and in accordance with the x-ray analysis they are probably pyrite or marcasite. At least three other minerals are also present in this area. A phase that appears fine-grained in the SEM contains predominately Ca, magnesium (Mg), C, and oxygen (O) and small amounts of Fe, and is most likely a dolomite. The large, circular grains in the SEM contain predominately Al, silicon (Si), sodium (Na), and O and is most likely analcite, but may be albite, which was shown also by x-ray analysis to be a constituent in this shale. Lastly, potassium (K) is associated with Al in a fine-grained silicate phase which may be a potassium clay or feldspar.

Area B was observed under the microscope as a tan-colored strip located on either side of the crack at $\sim$ 13,000 μm on the photomicrograph in Figure 1. This area is characterized by a significantly higher Fe concentration than in the darker areas of the sample. The Fe is associated with Mg, C, and O and lesser amounts of Ca, and occurs probably as Fe-rich dolomite or magnesite. Between grains of this material is a mineral that contains Al, Si, K, and O and may be a K-clay or feldspar. Another mineral phase contains Al, Si, and O, but no K or Na, and is possibly kaolinite although this mineral was not observed by x-ray diffraction. Quartz (SiO_2) and titanium dioxide (TiO_2) are also present in minor amounts in this area. X-ray diffusion studies did not identify the TiO_2 mineral.

Area C contains a black strip which is predominantly C, but also with significant amounts of Na, S, and O. The S is probably organic, as indicated by the shape of the $K\beta$ x-ray bands (7). A dolomitic region, containing less Fe and Mg than found in area B, and a silicate phase, containing Al and K, are distributed heterogeneously around the black strip. Small inclusions, generally less than 5 μm in diameter, contain Ti (titanium) and O and appear outside of the black strip.

Smith has suggested that the Green River shales were formed at pH levels between 8.5 and 10 (3). The carbonate types typically encountered under these conditions are calcite, dolomite, magnesite–siderite co-crystals and siderite. All these species were observed with the microprobe or by x-ray diffraction in the carbonate phases of the shale sample that we studied. Smith also reasoned that the presence of organic matter in the Green River shales should enhance the precipitation of silicate minerals. Our data is in accord with this postulation. Thus, we find that the organic-rich bands in this shale sample have higher concentrations of siliceous minerals than do the organic-deficient bands. These siliceous minerals occur as quartz and aluminosilicates.

Relatively large negative Eh values (reducing conditions) also have been predicted for the formation of the Green River shales (3). The observation that none of the sulfur present in shale sample studied is observed as sulfate agrees with this. The S is observed generally as FeS_2 particles less than 50 μm in diameter, most often observed in or adjacent to the organic-rich areas. Some organic S is observed in organic material.

Several other minerals were observed in this sample of Green River shale, include several phosphorus (P)-containing minerals, possibly in the form of apatite or fluoroapatite. Certain rare earth elements—cerium (Ce), lanthanum (La), praseodymium (Pr)—were found in some of the calcium phosphate regions. Phosphorus has ben detected as a trace constituent of Green River shales, and has been postulated to exist as apatite although it has not been identified by x-ray diffraction (8).

In contrast to the Green River shale, the Devonian shale is regarded generally as being elastic in origin. Mineralogically, our sample was composed primarily of quartz and K-clays or micas. Minor amounts of pyrite and alumina were present as well as trace amounts of siderite, kaolinite, dolomite, and marcasite. The siderite content was higher in this Devonian shale sample than in the Green River shale sample studied.

The Devonian shale sample generally had a darker appearance with fewer and less distinct bands than the Green River shale sample, and the organic matter appeared to be more evenly distributed over the 2.5-cm sample. Several areas of this sample were examined using the slow rastering microprobe techniques to determine the elemental distribution in this material. The observations made in two of the areas are presented below.

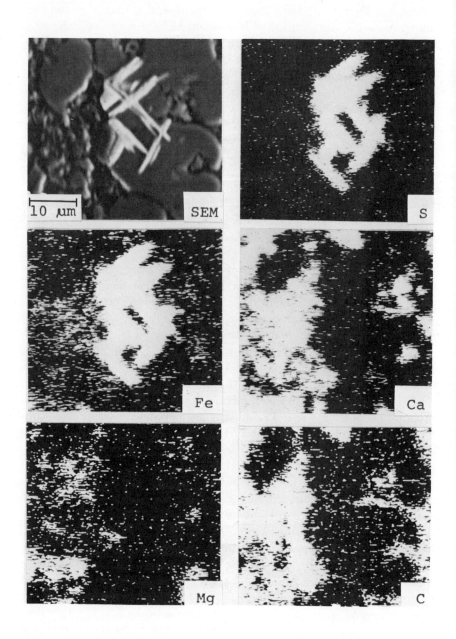

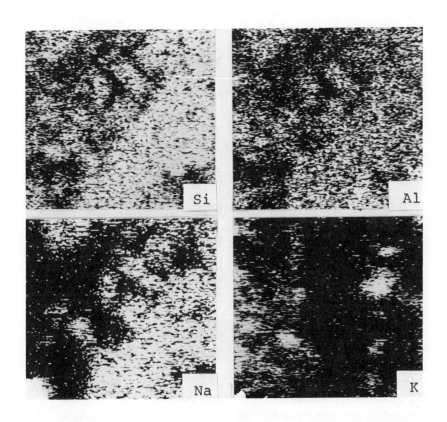

Figure 2. Scanning electron micrograph (SEM) and S, Fe, Ca, Mg, C. Si, Al, Na, and K x-ray intensity distributions in dark band of Mahogany zone shale

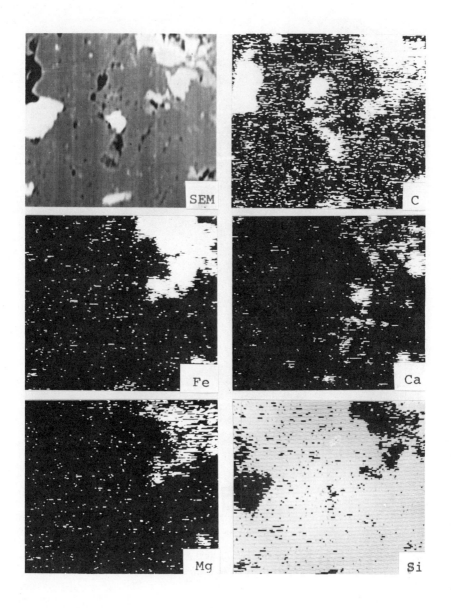

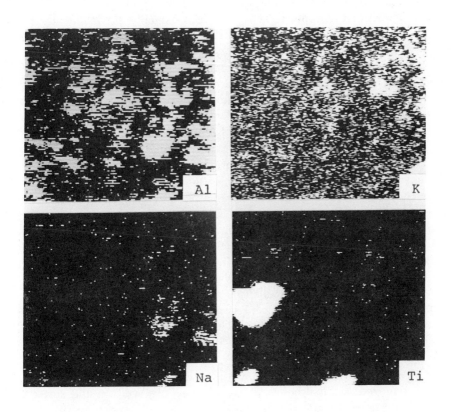

Figure 3. SEM and C, Fe, Ca, Mg, Si, Al, K, Na, and Ti x-ray intensity distributions in dark band of Devonian shale

One of the areas studied is shown in Figure 3 and is located in a darker, organic-rich band. The matrix of this area contains Si, primarily as quartz, but also as aluminosilicates with low concentrations of K. Two distinct carbonate regions can be differentiated in this sample. The first, shown in the upper right of the SEM, contains predominantly Fe with lesser amounts of Mg and even smaller amounts of Ca. Based on the x-ray analysis, this region probably is composed of siderite or Fe-rich magnesite. The second carbonate region located in the lower right of the SEM contains predominantly Ca and lesser amounts of Fe and Mg, and most likely is impure calcite although this mineral was not observed in the x-ray diffraction analysis. Several irregularly shaped particles shown in the lower left and center of the SEM contain Ti and O.

Another area studied in the Devonian shale sample is located also in a darker part of the shale structure (Figure 4). This area contained numerous, metallic-appearing particles as well as a black, irregularly shaped, carbon-rich region. The metallic particles contained only Fe and S and are mainly pyrite. Pyrite has been reported as the primary heavy mineral identified petrographically in a series of core samples from the Devonian shales of West Virginia (6). The black, irregularly shaped, organic region is surrounded generally by siliceous material consisting of quartz and aluminosilicates that contain only minor concentrations of K. As with the Green River shale, aluminosilicates containing K and Na associated together were not observed. Calcium (Ca), Mg, or Fe were not found in the aluminosilicates. This could suggest authigenic formation, but confirmation would require further studies, such as observation of cathodoluminescence (9).

After polishing, this sample was covered with a cap. When the cap was removed, small droplets of liquid were observed on the surface. Upon exposure to dry air, the liquid disappeared leaving small cubic crystals containing only Na and chlorine (Cl). Presumably, Na and Cl ions were leached from the interior of the sample and crystallized as NaCl on the surface as the liquid evaporated. Similar results were not observed with the Green River shale, suggesting that NaCl was not present in significant amounts.

This study has shown that the sample of Devonian shale is composed primarily of silicates with much lower amounts of carbonate minerals. Here, as with the Green River shale sample, the silicate minerals were associated primarily with the organic-rich areas of the shale, and when present, the carbonate minerals were found mainly in the organic-poor areas. The siliceous minerals of this Devonian shale sample were found to be quartz, illite, and muscovite, with trace amounts of kaolinite. Calcite, dolomite, Fe-rich dolomite or magnesite, and siderite were observed in the carbonate regions of the Devonian shale. However, the

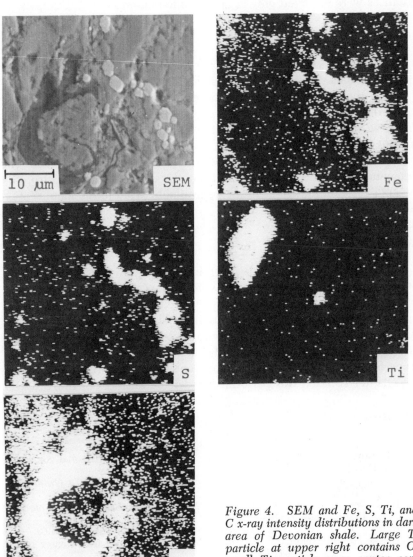

Figure 4. *SEM and Fe, S, Ti, and C x-ray intensity distributions in dark area of Devonian shale. Large Ti particle at upper right contains C; small Ti particle near center contains O.*

Devonian shale had regions with much higher concentrations of siderite than were found in the Green River shale sample. As in the Green River shale, the FeS_2 particles were less than 50 μm in diameter and in most instances were observed only in the organic-rich areas of the samples.

Summary

This paper is not intended to be a definitive microprobe comparison of the two shales, but, rather, is a preliminary study intended to demonstrate to those not familiar with the electron microprobe the utility of using this technique in studying energy-bearing carbonaceous shales. We have attempted to show that the microprobe can produce a large amount of information about the elemental associations and micromineralogy of these shales in a relatively short period of time. The elemental analyses reported here from this study are qualitative, but methods for quantitative analyses are available to aid in the identification of specific minerals (10).

Acknowledgment

We thank the Laramie (LERC) and Morgantown (MERC) Energy Research Centers for providing us with shale samples, as well as J. W. Smith of LERC for pointing out the need for these characterization studies and for his encouragement and helpful discussions. We also thank R. B. Roof of LASL for his help in determining the x-ray identification of the minerals in this study, and R. D. Reiswig of LASL for preparing the samples for microprobe analyses. The work upon which this chapter is based was performed under the auspices of the USDOE.

Literature Cited

1. Atwood, M. T., *Chem. Technol.* (1973) 617–621.
2. Overbey, W. K., Morgantown Energy Research Center, personal communication (1975).
3. Smith, J. W., *Rocky Mountain Association of Geologists, 1974 Guidebook,* pp. 71–79.
4. Robb, W. A., Smith, J. W., *Rocky Mountain Association of Geologists, 1974 Guidebook,* pp. 91–100.
5. Bradley, W. H., *U. S. Geol. Survey* (1929) Prof. Paper **158-E**, 87–110.
6. Larese, R. E., Heald, M. T., Morgantown Energy Research Center report MERC/CR–77/6 (1977).
7. Hurley, R. G., White, E. W., *Anal. Chem.* (1974) **46**, 2234–2237.
8. Desborough, G. A., Pitman, J. K., Huffman, G., *Chem. Geol.* (1976) **17**, 13–26.
9. Kastner, M., *Am. Mineral.* (1971) **56**, 1403–1442.
10. Bence, A. E., Albee, A. L., *J. Geol.* (1968) **76**, 382–403.

RECEIVED August 5, 1977.

Analysis of Oil Shale Materials for Element Balance Studies

THOMAS R. WILDEMAN

Department of Chemistry and Geochemistry, Colorado School of Mines, Golden, CO 80401

ROBERT R. MEGLEN

Environmental Trace Substances Research Program, University of Colorado, Boulder, CO 80309

A Fischer assay simulates the conversion of oil shale to usable fuels in an above-ground retort. The results of an extensive program of chemical analysis of major and trace elements in spent shale, oil, and water collected from the Fischer assay of a standard oil shale are presented. The concentration of major elements in raw and spent shale can be determined only to ±10% in this study. Two criteria show that fluorine and zinc may have been mobilized during the assays. The concentrations of arsenic and selenium in the Fischer assay retort water exceed the maximum permissible concentrations for drinking water.

Oil shale technology in the United States has an interesting history. The prospects for this resource have been known for over 100 years and research on the production of oil and gas from this type of rock has been conducted since the resource was first discovered. However, a commercial retort that would produce synthetic fossil fuel products in amounts comparable with the traditional sources of these products has never been built in the United States. The current technological interest in oil shale is the third time this resource has had such an intensive appraisal. Because of energy and mineral shortages, the prospects for development appear favorable. However, during the current appraisal, environmental considerations have been added to the questions to be answered before this new industry is developed. Consequently, recovery of fuel from oil shale is one of a few industrial processes which permits

0-8412-0395-4/78/33-170-195$05.00/1

the simultaneous consideration of process optimization and environmental factors at the engineering design stage. This will help to protect the surroundings and to ensure the maximum recovery of products from the feedstock. Before this can be achieved, elemental analyses of the raw shale and process products have to be performed so that the pathways of the inorganic and organic constituents produced can be determined. This paper describes such an elemental study. The retorting in this work was done by Fischer assay. Other studies of this nature have been performed recently on the ERDA–Laramie Energy Research Center's (LERC) controlled state retort and the Lawrence Livermore Laboratory's 125-kg retort (1, 2). TOSCO also has studied the distribution of some trace elements using Fischer assay products (3).

Properties of Green River Shale

Green River shale was deposited in a carbonate-rich salt water lake or evaporite environment of high pH (4, 5, 6). Because of this environment, the major minerals in the oil-rich Parachute Creek Member of the Green River Formation are quartz, dolomite, calcite, and feldspar. Clay is usually a minor mineral (7, 8). These properties make oil shale different from the typical marine shale in which clay is the predominant mineral. The environment of deposition affects the organic matter in oil shale because it is derived from algae, spores, and pollen. This sapropelic type of material produces organic constituents that have higher hydrogen-to-carbon ratios than other black shales or coals (6). This organic material also has lower amounts of humic acids which may retain elements such as uranium, and less sulfur compounds which can decompose to form inorganic sulfides (4).

The deposition from a saline lake of pH 8–10 tends to produce higher abundances of trace elements like molybdenum, strontium, boron, and lithium in oil shale because they can be soluble in such solutions (9, 10). In contrast, a typical marine black shale which has larger concentrations of humic material and sulfur compounds tends to have a greater abundance of zinc, cadmium, arsenic, lead, molybdenum, and uranium (9, 11). Of the trace elements analyzed to date, lithium, boron, arsenic, selenium, molybdenum, and antimony have abundances that are usually greater than ten times the average abundance of the earth's crust (2, 9, 10). In summary, the Green River shale has an organic and mineral composition that is different from marine shales, and has trace element concentrations that are typical of an average sediment.

These properties of oil shale lead to basic assumptions concerning the fate of trace elements during oil shale retorting. From an economic viewpoint, no trace elements occur in extractable quantities in typical oil shale. Any by-product recovery of inorganic materials probably will rely on the unusual minerals present such as dawsonite and nacholite (*6, 12*). From an environmental viewpoint, the elements which may be most harmful are those in high abundance in oil shale and those which are mobile in a basic environment. Our research program is concentrating on the study of boron, fluorine, arsenic, selenium, and molybdenum. Although lead, cadmium, and mercury are not in unusually high concentrations in oil shale (*1, 2, 9, 10, 13*), the retorting process may mobilize them so that they go to retort product other than the spent shale. Mercury has been identified already as moving from the solids and accumulating in the liquid and gaseous products (*1, 2, 13*). The purpose of our studies is to see if these assumptions about trace elements in oil shale are true.

Comments on Fischer Assay Analyses

Heistand (*14*) has described the Fischer assay as a bench-scale performance test which gives a measure of the quantity of liquid hydrocarbons that can be retorted from a solid sample. The procedure used in this study is the TOSCO Modified Fischer Assay which is one of the acknowledged testing methods (*14, 15*). In this procedure, 100 g of −65 mesh material is weighed in an aluminum cup and placed in a stainless steel retort. A controlled pyrolysis temperature program lasting for 72.5 min is used. This includes a 20-min soak period at the maximum temperature of 500°C. Liquid products of the retorting are collected in a condenser and receiver both cooled by ice water. In this study, gases not condensed at 0°C were allowed to escape; however, provisions can be made for their collection (*15*). By determining the mass of oil, water, and spent shale produced, the oil yield can be calculated (*15, 16*). Two aspects of Fischer assays render the test inadequate as a standard analytical procedure: it appears that the oil yields vary depending on the procedure used and the assay is subject to uncontrollable variables in technique which adversely affect the precision (*14*). Some of the variables leading to poor reproducibility are the size to which the sample is ground, the uniformity of the grain size, the degree to which the fine particles are retained, the diligence used in controlling the retort temperature program, and the temperature at which the receiver is maintained (*14, 15*).

The oil yield results of the Fischer assay obtained by the modified TOSCO procedure have been discussed previously (*16*) and are summarized in Table I. The relative standard deviation in this set of assays was 2.5% on nine oil-yield assays. TOSCO has obtained precision limits of 0.6%. Further examination of the weight fractions in 46 retort runs showed that the major source of this scatter was in the weight of liquid product obtained. This probably can be attributed to the inadequate control of the retort temperature program. The above summary indicates

Table I. Results of Fischer Assay Analyses on the Standard Shale[a]

	Value	Relative Standard Deviation %	No. of Analyses
Oil yield (gal/ton)	39.8	2.5	9
Oil specific gravity	0.909	0.4	9
Raw oil shale wt (g)	100.0	0.1	120
Spent oil shale wt (g)	80.4	0.4	46
Liquid wt (g)	16.3	2.6	46
Oil wt (g)	14.9	2.6	46
Water wt (g)	1.4	8.0	46

[a] Refs. 15 and 16.

that the Fischer assay has limitations. However, in the absence of any adequate alternative, it probably will continue to be used as a standard bench retort test.

While the Fischer assay is plagued by variations, the sources of variability encountered in sampling a pilot plant or full scale production retort are even greater. The feedstock is much less uniform and the temperature profile of the retorting chamber may be subject to large variations. The Fischer assay was chosen for the mass balance study because it is likely to become a standard bench test and it offers the best prospects for control of retorting parameters. The Fischer assay is similar to above-ground retorts where the shale is heated indirectly such as the TOSCO II or Paraho Indirect Process (17). In the Fischer assay, the liquid product is distilled from the shale; whereas in situ retorting permits liquid oil and water to remain in contact with unretorted shale and this will raise the elemental abundances in these liquids (18, 19). Also, the Fischer assay does not introduce combustion gases as do the direct combustion and in situ retort processes. This may change the oxidation state of some elements such as arsenic or mercury and aid in the transfer of some of the volatile elements from the retort (17, 19).

Analytical Procedures

Oil shale and its retort products are a challenge to the trace element analyst. The rock and spent shale contain silicates, both low and high molecular weight organics, and inorganic sulfides. This mixture requires that great care be taken during sampling, preparation, and digestion. The oil has low concentrations of the trace elements. It typically has a pour point of ca. 25°C so that sampling, handling, and digestion pose special problems. The water is quite unusual (18); the primary constituents are NH_4^+ and HCO_3^- and dissolved organics which are stable in a

basic solution. It also requires special handling procedures. The techniques which we have found useful in the handling and analysis of these
materials for trace elements are summarized in the following sections.

Preparation and Storage. Much of the preparation work on the
standard shale and spent shales has been reported previously (16, 20).
The raw shale was taken from a 3-ton pile of −1/2 mesh feedstock
which was mined from a fresh face of the Dow Mine of the Colony
Development Company at the head of Parachute Creek in the Piceance
Creek Basin, Colorado. TOSCO II pilot plant operations produced a
yield of 37–38 gal/ton. The shale is designated as OS-1; SS-2 is a spent
shale from the pilot plant produced from the same feedstock.

No problems were encountered in crushing the samples, however
the −10 mesh raw shale particles are difficult to pulverize because of
their resinous nature. A shatter box with hardened steel or tungsten
carbide grinding chambers gave good results. The concentrations of
cobalt, copper, zinc, and lead are low enough that contamination from
preparation equipment may be a problem (1, 16, 20). Therefore, sieves
and disc pulverizers were avoided and hardened steel and tungsten carbide were used to minimize contamination. Spent shale (SS) is typically
more friable than raw shale and it grinds easily, but this renders it
extremely dusty. Samples were stored in a refrigerator.

Since oil shale is fine-grained and typically does not contain any
trace element-loaded minor minerals such as allanite or zircon, inhomo-
geneity of the pulverized sample is not likely. Table II shows analytical
results for rubidium and strontium on different 0.500-g splits (16) and
300-μg replicates (21). For most of the elements studied, homogeneity
is retained down to the 300-μg sample sizes.

Retort water collection and preservation poses special problems because of its unusual chemical characteristics. For example, the electro-
chemical properties of the Fischer assay water collected in this study
were pH = 8.95, conductivity = 29,000 μmho/cm at 25°C, and Eh =
−310 mv. The major species in the water are NH_4^+ and HCO_3^-. Con-

**Table II. Results of Homogeneity Tests for Rubidium (Rb) and
Strontium (Sr) on the Standard Shale and a Spent Shale**

	Raw Shale		Spent Shale	
Target Size	0.500 g	300 μg	0.500 g	300 μg
Number of Samples	8	6	8	6
Average for Sr (ppm)	584	584	771	735
Standard deviation for Sr (ppm)	24	90	17	23
Relative standard deviation for Sr (%)	4.1	15	2.2	3.1
Average for Rb (ppm)	60.1	63.3	80.7	93.4
Standard deviation for Rb (ppm)	2.7	1.9	1.5	4.4
Relative standard deviation for Rb (%)	4.5	3.0	1.9	4.8

ventional methods of preserving water are suspect for this sample (22). Acidifying to pH less than 3 requires large amounts of HNO_3 and causes the separation of organic acids and sulfur (23) which render the sample heterogeneous. Our analyses of acidified water have shown losses in arsenic and selenium by this procedure. Since vacuum or pressure filtration may alter the sample, gravity filtration was used to remove oil and large particles. The water then was bubbled with N_2 to remove any oxidizing gases and placed in Teflon bottles for refrigerated storage. There was no visible change after one year for water prepared this way.

The Fischer assay oil was free of particles. Little residue was found upon filtering a 500-mL sample and the differences in elemental analyses between the filtered and unfiltered oil were within analytical error. However, this may not be the case for oil from other retorts and filtration may be necessary. Nitrogen gas was bubbled through the samples and they were stored in Teflon containers in a refrigerator.

Study of the recent publication by the National Bureau of Standards on sampling, sample handling, and analysis (24) suggests that the best method of handling the oil and water is to filter, bubble with N_2, and not acidify. Storage in acid-washed, conventional polyethylene bottles in a freezer is recommended.

Problems in Analyzing Samples. The samples encountered in this study fall into three general categories: geological (oil shale and spent shale), waters (process waters), and petroleum (shale oil). The chemical treatments required for these materials have been adapted from conventional methods commonly used for the analysis of similar materials. However, analysis of the oil shales has presented several problems not common to other geological materials.

The high organic content of these shales complicate the conventional basic salt fusion techniques which are supposed to render the silicates soluble. The temperatures required for these fusions are so high that the organic constituents and volatile trace elements are lost during the heating process. This occurs at a temperature below the nominal fusion temperature. Under these circumstances any fixing properties which the fusion flux has are not effective. This problem is particularly troublesome when arsenic and selenium are to be determined. Many inorganic salts of these elements are volatile at temperatures below the temperatures at which complete fusion is effected. In addition, the presence of organic material enhances the loss through volatilization of these elements. Accurate trace element analysis of these materials requires a treatment which is sufficiently vigorous to ensure complete decomposition of the silicate matrix, and the treatment also must provide for the retention of the volatile components.

Strong acid digestions of geological materials have been used effectively for the analysis of some elements. These procedures are performed usually at temperatures below most fusions. Under carefully controlled conditions loss of volatile components can be minimized. However, these methods do not decompose the silicate matrix completely. Therefore, some of the elements remain bound in the insoluble silicate residue. Chemical analysis of the resulting solutions lead to low results for many elements.

Because of the difficulties outlined here, we have had to adopt the tedious procedure of performing several different treatments on each sample. The treatments used in this lab have been selected on the basis of their applicability to a particular trace element or suite of elements. To date no single sample decomposition technique has been effective for all of the target elements of this study.

The analysis of oil shale samples for arsenic and selenium present special problems mentioned earlier. Fusion of the shale with sodium peroxide at 600°C in zirconium crucibles has been used with 90–95% recovery. The advantage in this method is that the fusion is effective at a temperature below those of most other basic-flux techniques. More importantly, however, the sodium peroxide maintains oxidizing conditions which minimize loss through volatilization.

While acid digestions do not competely decompose the silicate matrix of these shales, it has been possible to effect 90–95% removal of arsenic and selenium from these samples by digestion with a sulfuric-nitric-perchloric acid mixture. However, this method has stringent requirements. Oxidizing conditions must be maintained at all times (the presence of perchloric acid or hydrogen peroxide is necessary). The temperature must be controlled carefully and conditions for a slow reflux must be provided. The reflux capability provided by the long necks of Kjeldahl flasks helps to minimize volatilization of arsenic and selenium compounds. A similar digestion procedure is used for oils. National Bureau of Standards coal and fly ash (SRM 1632, 1633), process waters, and plants have been analyzed successfully using this technique.

Fusion Methods. Na_2CO_3 for major elements: pulverized geological material (0.5 g) is mixed with 2 g of reagent-grade Na_2CO_3. The mixture is placed in a covered 25-mL platinum crucible. The crucible is placed in a 200°C oven and the temperature is raised to 1000°C over a period of 2 hr. After the mixtures have been maintained at 1000°C for 0.5 hr the fusions are cooled, removed, and dissolved in 3% HNO_3. Appropriate volumes and dilutions are dictated by the method of analysis used. Filtration may be necessary. A similar procedure using $LiBO_2$ flux has been used also. Na_2CO_3 for boron: the same conditions used for the major element fusions are used when boron is to be determined. However, no borosilicate glassware is used for storage, dilution, or additions. Polyethylene labware is used and 6% (v/v) H_2SO_4 is substituted for HNO_3. Boron is determined spectrophotometrically using Azomethine H (25).

NaOH for fluorine: Approximately 0.5 g of pulverized sample is placed in a 100-mL nickel crucible. The sample is moistened with distilled water and 6 mL of 16M NaOH solution is added. The mixture is dried on a hot plate at 150°C. The mixture is then fused for 0.5 hr in an oven at 600°C. The fusion cake then is dissolved by the slow addition of concentrated HCl. The pH of the solution is adjusted to $\sim$ 8. Monitoring of the pH is accomplished with pH test paper. The sample is cooled and filterd through Whatman No. 40 filter paper and transferred to a 100-mL volumetric flask and diluted to the mark. The sample then is stored in a polyethylene bottle. Analysis is performed with a Fluoride Selective electrode using a total ionic strength adjustment buffer. Further details of the method are described by McQuaker and Gurney (26).

Na$_2$O$_2$ for arsenic and selenium: Approximately 0.5 g of pulverized samples are mixed with 2 g of Na$_2$O$_2$ in a 30-mL zirconium crucible. The covered crucibles are placed in a 450°C oven and the temperature is raised to 600°C and maintained for 0.5 hr. The fusion is dissolved in 6M HCl. (This acid concentration is convenient for the subsequent analysis by hydride generation coupled with atomic absorption spectrophotometry (A.A.S.)). The final volume of these solutions is 100 mL. Separate aliquots with and without NaI reducing agent are taken for arsenic and selenium analysis, respectively.

Acid Digestions. Oil and process waters: The procedure used has been described by Walker, Runnels, and Merryfield (27). In some cases, use of perchloric acid is undesirable. Slow addition of 30% H$_2$O$_2$ may be substituted for perchloric acid. Several additions may be necessary to effect complete destruction of organic matter.

Shales for arsenic and selenium: The same procedure described for oil is used. However, the water-cooled distillation head is not used.

Instrumental Methods of Analysis. The problems encountered in fusing and digesting oil shale make instrumental methods attractive. Instrumental neutron activation analysis (INAA) and x-ray fluorescence (XRF) analysis either in the wavelength dispersive or energy dispersive mode are especially attractive since they are capable of accurate multi-element analyses.

When INAA is used and the concentration of Na in the sample is above 1%, the Na^{24} γ-rays mask other useful γ-ray peaks. In particular, arsenic and zinc abundances are not as easily determined in the shales as they are in coal by INAA (28). The oil and water appear to present no leakage problems at fluxes of 2×10^{12} n cm^{-2} sec^{-1} for irradiation periods of 2 hr. These liquids were irradiated in sealed polyethylene vials inside of sealed polyethylene bags. The neutron activation analyses were performed at the U.S. Geological Survey in Denver, Colorado.

The analysis of the elements from manganese through molybdenum using low-current energy dispersive XRF yields reasonable results (21, 29). The concentration of lead is high enough that the L-lines of lead interfere with the K-lines of arsenic and selenium. Corrections have to be made in these cases. The analysis of the oil and water present special problems in target preparation. We currently are investigating the addition of an internal elemental spike such as chromium or rhodium to these samples so that the analyses can be done without digesting or evaporating the samples (30). The XRF analyses were performed at the University of Colorado in Boulder, Colorado.

Major Element Results

The focus of this research and other mass balance studies has been on trace elements (1, 2, 3). However, in future studies on speciation it will be necessary to know the concentrations of the elements present in amounts above 1%. Therefore, analyses of the oil shale and spent shale samples were performed for these elements. Atomic absorption and colorimetry were used for many of these analyses. Some major element results also were obtained by the broad-range instrumental analysis surveys. The comparison of the results obtained by the different techniques shows large discrepancies.

Table III is a summary of the analytical data for potassium and iron for the raw shale (OS-1). In all instances at least two samples were ana-

Table III. Results of Analyses for Potassium (K) and Iron (Fe) by Different Methods in Different Laboratories on Oil Shale OS-1[a]

	K		*Fe*	
	Average	*Standard Deviation*	*Average*	*Standard Deviation*
Atomic absorption	1.17	0.09	2.42	0.18
Atomic absorption standard additions	1.70	0.08	—	
Instrumental neutron activation analysis	1.33	0.06	1.88	0.07
Wavelength dispersive x-ray fluorescence	0.80	0.01	1.52	0.06
Energy dispersive x-ray fluorescence	1.56	0.10	1.47	0.15

[a] All values are in %.

lyzed. All the analysts have had experience in analyzing geologic materials. The conclusion is that the concentration of potassium and iron in this oil shale is not known to within ±10%. This typically is considered as unacceptable for elements that account for a significant portion of the specimen. Results on spent shales and on other major elements show the same poor reproducibility. The values of the standard deviation for each method indicate that the precision is reasonable. Furthermore, analyses on oil shale samples, including OS-1, have been performed by Lawrence Berkeley Laboratory by INAA and energy dispersive XRF (29). The major element results obtained by both methods are within 5%. Apparently oil shale and spent shale are sufficiently different from other geochemical specimens that the matrix is presenting unexpected analytical problems.

The association in oil shale of organic material with inorganic minerals makes the matrix different. If this is the case, methods of analysis of major elements that require a chemical attack may be questionable. In this regard, the nondestructive techniques, such as XRF and INAA, used for trace element analyses may be also useful for establishing the concentrations of major elements. Interlaboratory analyses of a standard oil shale by the broad range instrumental methods are needed to establish accuracy and to evaluate methods. The analysis of other geologic standards will not resolve the situation. Such a program of round-robin analyses of standard oil shales is being carried out now by Battelle Northwest Laboratories, Lawrence Berkeley Laboratory, and the Environmental Trace Substances Research Program of Colorado (31). For the major elements, an outcome of these interlaboratory analyses will be the establishment of accurate methods of analysis for aluminum in oil shale. Plans are being considered to extract alumina from the dawsonite contained in oil shale (12), and reliable aluminum analyses are needed for realistic assessments.

Table IV.   Trace Element Concentrations in Oil

| | | TOSCO | |
Element	Method	Raw Shale OS-1	Spent Shale SS-2
Boron	Color	126	146
Fluorine	Elec.	1070	1120
Sodium	AA		
Potassium	XRF		
Calcium	XRF		
Manganese	XRF	196	240
Iron	XRF		
Nickel	XRF	24	31
Copper	XRF	44	52
Zinc	XRF	70	82
Arsenic	XRF	64	80
Arsenic	AA	75	86
Selenium	XRF	2.9	3.2
Selenium	Fluor	3.5	4.4
Rubidium	XRF	60	81
Strontium	XRF	580	770
Yttrium	XRF	7.7	10.6
Molybdenum	XFR	28	38
Molybdenum	Color	26	30
Lead	XRF	27	39
Uranium	NAA	5.4	6.6

[a] All values are in ppm.

Trace Element Results

The analytical results for trace elements appear better than those for the major elements. Table IV is a summary of results for the raw shale (OS-1), and the spent shale from the TOSCO II pilot plant (SS-2), and the spent shale (FS), unacidified water and unfiltered oil from the Fischer assay studies. The spent shale SS-2 is the residue from the retorting of the same raw feedstock from which OS-1 was taken. In the case of the solid samples, the concentrations are from multiple analyses of different splits. In all cases, the relative deviations of multiple analyses lie within ±10%. Comparison of these results with those from other laboratories on different splits of the same solid samples show agreement within ±2σ (*31*). The two exceptions are manganese and zinc for which the results reported here are low in comparison with other methods. Analyses of NBS standard coal (SRM 1632) and coal fly ash (SRM 1633) are included also in Table IV. The concentrations in raw and spent shale are similar to those reported by TOSCO except for selenium where their values range from 10–16 ppm (*3*).

Shale Materials and in NBS Coal and Coal Fly Ash[a]

Fischer Assay			NBS	
Spent Shale FS	Oil	Water	Coal 1632	Fly Ash 1633
145	NA	2.8		
1080	< 3	1.7		
	17	43		
		20		
		1.4		
224			16	400
		1.0		
29	0.6	2.3	14	98
47			16	106
72	0.5	0.2	18	185
79	18	4.2	5.0	62
87	26	0.9		
2.9	0.8	3.1	2.4	6.8
4.3	0.8	0.03	2.9	9.6
86			19	130
770		0.06	152	1490
10.2			7.4	66
36		0.13	1.7	23
28	ND	ND		
37			6.0	66
6.4	NA	NA	1.6	

Although most of the elements have been determined by XRF (*21*), some other methods were used. The fluorometric method for selenium uses diaminonaphthalene (*32*). The colorimetric method for molybdenum uses potassium thiocyanate (*33*). The uranium analyses were done by delayed neutron activation analysis (*34*). For the XRF analyses of the oil and water, a blank value implies that there were no x-rays above background for that element. Two elements conspicuously missing from Table IV are cadmium and mercury. Preliminary analyses for these two elements have not yielded reproducible results. Further work is needed before we can make definitive statements about cadmium and mercury.

Discussion of Results

From an environmental viewpoint, the mass balance analysis is aided by the calculation of two parameters (*1, 35*):

1. Enrichment Ratio (ER). This parameter is determined by dividing the elemental concentration in each output by that in the raw shale. This ratio expresses the relative enrichment in each output stream independent of the mass of the input and output stream.

2. Relative Imbalance (RI). This parameter is determined as a percent by the following equation:

$$RI = \frac{\text{mass in} - \text{mass out}}{\text{mass in}} \times 100$$

This percent measures the degree of closure obtained taking into account the mass of the respective element involved in the input and output streams.

In some instances, the RI value obtained is a large negative number. This is usually an indication that the retort process has added that respective element or that contamination occurred during sample handling. In this study, care was taken to minimize this problem. All the retorts and aluminum inserts used in this Fischer assay were scrubbed and rinsed with HNO_3 (*15*). All new glassware was used and was also rinsed with HNO_3 prior to use. All equipment was used in a Fischer assay of OS-1 before samples were collected for the mass balance studies.

The RI values were obtained using the masses of the raw shale feedstock and the oil, water, and spent shale products shown in Table I. The gas plus loss amounts to 3.3% in these Fischer assays. The gas was not analyzed for trace elements and was neglected in the calculation of RI.

Trace Elements in the Oil and Water. The enrichment ratios for the trace elements in the oil and water are far less than 1.0 in most cases. The ER for arsenic in oil is 0.3. The ER for selenium in oil is 0.25; for selenium in water, the ER is 1.0 if the XRF value is used. Most elemental concentrations in the liquid products are negligible in the mass balance calculations. For the calculation of RI for each element, only selenium and arsenic in the oil contribute as much as 5% to the balance closure. These results are typical of what others have found (1, 3). Only mercury in oil has an ER that is greater than unity (1). The trace elements in the Fischer assay oil are similar to the results obtained for other shale oils (1, 2, 3). The Fischer assay water is similar to other shale process waters, except that the arsenic and selenium concentrations obtained by the XRF method appear high (1, 2, 3, 18).

The concentration of arsenic in shale oil is higher than other crude oils (3). However, this will have to be removed before refining because it might poison the catalysts used in most modern refinery processes.

If the retort water is considered for domestic or irrigation uses, it would have to be treated to remove a number of contaminants. Of the major constituents in retort water, the NH_4^+, HCO_3^-, and organic compounds in the water clearly make it unsuitable for other uses (18). Of the trace constituents, the arsenic and selenium concentrations listed in Table IV are above the maximum permissible concentrations for drinking water (36). The boron concentration may make water unsuitable for irrigation (22). Other studies have found silver and lead concentrations in retort waters have exceeded the maximum permissible concentrations for drinking water (1, 2, 36). Numerous studies for the treatment of the retort water have been initiated (37). The objectives of these studies are usually to find a method to recover the ammonia and organic material from the water so that treatment costs will be lowered through by-product recovery.

Mass Balance Considerations. The values of ER for the Fischer assay spent shale are contained in Table V. If it is assumed that the relative standard deviation in the analyses is ±10%, then the relative probable error in ER would be ±14% if the analytical errors were indeterminant and ±20% if the errors were determinant (38). The mass ratio of OS-1/FS is 1.24 as derived from the assay data in Table I. It is not possible to conclude that any trace elements are mobilized from the solid material during the assay retorting. The ER results obtained for arsenic, selenium, and molybdenum indicate the importance of analytical precision in detecting any trace element mobilization during oil shale retorting. The values of RI contained in Table V show a similar dependence on analytical precision. The probable errors in these values are also between 14 and 20% if the relative standard deviation in the analytical results is assumed to be ±10%. These results indicate that, within experimental error, none of the trace elements have been lost during Fischer assay. More definitive conclusions on whether elements are mobilized or lost can only be reached with more precise analytical

Table V.   Enrichment Ratios (ER) and Relative Imbalances (RI)
for Trace Elements in the Fischer Assay of Oil Shale

Element	ER^a Fischer Shale	RI in %
Boron	1.15	+ 7
Fluorine	1.00	+19
Manganese	1.14	+ 8
Nickel	1.21	+ 3
Copper	1.07	+14
Zinc	1.03	+17
Arsenic	1.23	− 4
Arsenic ^{b}	1.16	0
Selenium	1.00	+16
Selenium ^{b}	.1.23	0
Rubidium	1.43	−15
Strontium	1.33	− 7
Yttrium	1.32	− 6
Molybdenum	1.29	− 3
Molybdenum ^{b}	1.08	+15
Lead	1.37	− 7
Uranium	1.19	+ 5

^{a} From Table I, the mass ratio of OS-1/FS = 1.24.
^{b} Arsenic analysis by atomic absorption; selenium analysis by fluorimetry; molybdenum analysis by colorimetry.

data. Mercury is the only trace element that is lost during oil shale retorting (1).

Considerably more effort will be required to improve the precision of the analytical methods. Therefore, it is useful to try to establish some further criteria by which to evaluate the mass balance results so that efforts can be focused on the elements that appear to be mobilized or lost. Based on the analytical results and the values of ER and RI, the following two criteria may be useful:

1. Typically, spent shale amounts to 80% or more of the mass for retort products. Assuming a value of 80% and equal concentrations of a trace element in the raw shale and spent shale and zero abundances in the oil and water, one would obtain an ER of 1.00 and an RI of +20%. These values are just at the limits of the probable error in ER and RI found here. Thus, a reasonable criteria for suspicion that an element has been mobilized from the solids during retorting is that the concentration of that element in the spent shale is at or below that found in the feedstock.

2. Lithophile elements such as magnesium, potassium, calcium, rubidium, strontium, the rare earth elements, and uranium (9) are not expected to be mobilized during oil shale retorting. Many of these elements can be determined quite precisely at trace element levels. Thus, investigation of the values of ER and RI for these elements should help to establish the reliability of the ER and RI values.

Using the first criterion, fluorine and zinc can be suspected of being mobilized during Fischer assay. Using the second criterion, the values of RI for lithophile elements range from a low of -15% for rubidium to a high of $+8\%$ for manganese. This suggests a probable error of $\pm15\%$ in the value of RI. This value for the probable error of RI also suggests that fluorine and zinc have been lost during Fischer assay. The apparent zinc loss may be reasonable since the Battelle group found 15,000 to 40,000 ng/m^3 of zinc in the off-gas from the LERC 10-ton retort (2). Most of this zinc was as vapor or particles below 0.45 μm in size. The apparent fluorine loss has not been confirmed on actual retorts, and the imbalance seen here may be the result of analytical uncertainty.

Nothing has been said of the spent shale SS-2 which was produced from the TOSCO II pilot plant. This is because it is difficult to say how representative the 50 lb of OS-1 is of the 100 tons of $-1/2$ in. mesh feedstock. TOSCO determined that the 100 tons assayed at 37–38 gal/ton of oil. The Fischer assays performed in this study yield a value of 39.8 gal/ton (16). The difference in oil yields is outside of analytical uncertainties (16). If OS-1 can be considered representative of the pilot plant feedstock, then the abundances of trace elements in SS-2 compared with those in OS-1 would give ER values greater than unity for all trace elements. The elements fluorine and zinc in SS-2 have ER values of 1.05 and 1.17, respectively. Based on these results, it appears that the trace elements for which data are reported here are not mobilized during the TOSCO II process.

Although there are differences between the two, it is useful to compare the mass balance studies done on oil shale retorts with the more familiar studies on coal-fired power plants. In coal burning, ER values of 5–10 are found for lead, arsenic, zinc, molybdenum and antimony in the fly ash particles that are collected in the stack scrubbers and in the air downwind from a plant (35, 39). Values of this range have not been found yet in oil shale retorting studies and there are reasons to believe that high enrichments won't be found. This is because the differences between an oil shale retort operation and a coal-fired power plant operation are large. First, the fly ash amounts to less than 10% of the coal feedstock, whereas the spent shale amounts to 80% of the retort feed. Secondly, the temperatures in a retort seldom exceed 650°C whereas those in a coal furnace reach 1500°C (17, 35, 39). Thirdly, oil shale is retorted under reducing conditions using as little input air flow as possible while coal is burned under more oxidizing conditions. Finally, an oil and water mist is formed in many retorts. This mist may function as a scrubbing agent in a manner similar to the aerosols formed in stack scrubbers in coal-fired power plants (17). All of these conditions act to make oil shale retorting less prone to trace element emissions than coal burning.

Fox and co-workers (1) have made an analysis of what species would have vapor pressures that would be high enough to allow escape under conditions of 500°–600°C in a reducing environment. Their study shows that elemental mercury, arsenic and cadmium, and HgS and As_2O_3 would have sufficient vapor pressures. Elemental zinc, if formed, would have a high enough pressure if the temperature rose to 650°C. The reducing conditions help to retain arsenic since As_2O_3 is more volatile than elemental arsenic.

The question of what species in raw shale contain those elements which might escape is also important to the question of what the eventual fate of these elements might be. Desborough and co-workers have begun to study this question (7, 10). Their results suggest that the majority of the arsenic is bound up in pyrite and that zinc occurs primarily as ZnS. Fluoride occurs as both fluorite and cryolite (Na_3AlF_6). Such observations can help to determine which trace elements are most likely to be released under oil shale retorting conditions. For example, Fox (1) observed that if the arsenic is tied up in pyrite then its release should not be expected since pyrite should be stable under oil shale retorting conditions. This implies that the arsenic that occurs in shale oil is partially present as organometallic arsenic in the raw shale.

In summary, it appears that oil-shale retorting is a reasonably clean process with regard to trace element loss from above-ground retorts. Careful control of the fate of the process water should be established. Possible releases will depend on how the trace elements are contained in the raw and spent shale and the temperature and reducing conditions under which the retort is operated. Further studies on raw and spent shale should address these questions. How the retort conditions could affect the emission of trace elements can begin to be seen in the results of the mass balance studies performed by the Berkeley group (1).

Acknowledgments

We thank the University of Colorado x-ray laboratory personnel for the fine work they did analyzing these samples. We thank the Berkeley oil shale group for the opportunity to study the results of their analyses prior to publication. The help of TOSCO Corporation in providing samples and use of their Fischer assay facilities also is appreciated. The reviews of P. Fox and G. Desborough substantially improved the manuscript. Finally, we thank the ERDA for their financial support of this study under Grant No. C00–4017–1.

Literature Cited

1. Fox, J. P., McLaughlin, R. D., Thomas, J. F., Poulson, R. E., *Oil Shale Symp. Proc. 10th*, K. B. Reubens, Ed., Colo. Sch. Mines, Golden, CO (1977) 223.
2. Fruchter, J. S., Laul, J. C., Petersen, M. R., Ryan, P. W., *Prepr., Div. Pet. Chem., Am. Chem. Soc.* (1977) **22**(2), 793.
3. Shendrikar, A. D., Faudel, G. B., *Environ. Sci. Technol.* (1978) **12**, 332.
4. Swanson, V. E., *U.S. Geol. Survey Prof. Pap.* (1960) **356-A**, 44.
5. Roehler, H. W., "Guidebook to the Energy Resources of the Piceance Creek Basin, Colorado," p. 57, Rocky Mountain Assoc. of Geologists, Denver, 1974.
6. Smith, J. W., "Guidebook to the Energy Resources of the Piceance Creek Basin, Colorado," p. 71, Rocky Mountain Assoc. of Geologists, Denver, 1974.
7. Desborough, G. A., Pitman, J. K., "Guidebook to the Energy Resources of the Piceance Creek Basin, Colorado," p. 81, Rocky Mountain Assoc. of Geologists, Denver, 1974.
8. Robb, W. A., Smith, J. W., "Guidebook to the Energy Resources of the Piceance Creek Basin, Colorado," p. 91, Rocky Mountain Assoc. of Geologists, Denver, 1974.
9. Mason, G., "Principles of Geochemistry," 3rd ed., p. 329, Wiley, New York, 1966.
10. Desborough, G. A., Pitman, J. K., Huffman, C., Jr., *Chem. Geol.* (1976) **17**, 13.
11. Vine, J. D., Tourtelot, E. B., *Econ. Geol.* (1970) **65**, 253.
12. Smith, J. W., Young, N. B., *Q. Colo. Sch. Mines* (1975) **70**(3), 57.
13. Donnell, J. R., Shaw, V. E., *J. Res. U.S. Geol. Surv.* (1977) **5**, 221.
14. Heistand, R. N., *Am. Chem. Soc., Div. Fuel Chem., Prepr.* (1976) **21**(6), 40.
15. Goodfellow, L., Atwood, M. T., *Q. Colo. Sch. Mines* (1974) **69**(2), 205.
16. Wildeman, T. R., *Prepr., Div. Pet. Chem., Am. Chem. Soc.* (1977) **22**(2), 760.
17. Prien, C. H., "Guidebook to the Energy Resources of the Piceance Creek Basin, Colorado," p. 141, Rocky Mountain Assoc. of Geologists, Denver, 1974.
18. Jackson, L. P., Poulson, R. E., Spedding, T. J., Phillips, T. E., Jensen, H. B., *Q. Colo. Sch. Mines* (1975) **70**(4), 95.
19. McCarthy, H. E., Cha, C. Y., *Q. Colo. Sch. Mines* (1976) **71**(4), 85.
20. Wildeman, T. R., "Preparation of Standard Oil-Shale Samples OS-1, SS-1, and SS-2," Nat. Bur. Standards Special Publication, in press.
21. Alfrey, A. C., Nunnelley, L. L., Rudolph, H., Smythe, W. R., "Advances in X-ray Analysis," Vol. 19, p. 497, Kendall/Hunt, Dubuque, Iowa, 1976.
22. Brown, E., Skougstad, M. W., Fishman, M. J., "Techniques of Water Resources Investigation," U.S. Geol. Survey, Book 5, Chap. A1, p. 160, 1970.
23. Stuber, H. A., Leenheer, J. A., *Am. Chem. Soc., Div. Fuel Chem., Prepr.* (1978) **23**(2), 165.
24. LeFleur, P. D., Ed., "Accuracy in Trace Analysis: Sampling, Sample Handling, Analysis," *Nat. Bur. Stand. (U.S.) Spec. Publ.* (1976) **422**, 1304.
25. John, M. K., Chauah, H. H., Neufeld, J. H., *Anal. Lett.* (1975) **8**, 559.
26. McQuaker, N. R., Gurney, M., *Anal. Chem.* (1977) **49**, 53.
27. Walker, H. H., Runnels, J. H., Merryfield, R., *Anal. Chem.* (1976) **48**, 2056.
28. Block, C., Dames, R., *Anal. Chim. Acta* (1973) **68**, 11.
29. Giauque, R. D., Garrett, R. B., Goda, L. Y., *Anal. Chem.* (1977) **49**, 62.
30. Kubo, H., Bernthaler, B., Wildeman, T. R., *Anal. Chem.*, in press.

31. Wildeman, T. R., Fox, J. P., Fruchter, J. S., in preparation (1978).
32. Chan, C. C. Y., *Anal. Chim. Acta* (1976) **82**, 213.
33. Ward, S. N., *Anal. Chem.* (1951) **23**, 788.
34. Stuckless, J. S., et al., *J. Res. U.S. Geol. Surv.* (1977) **5**, 83.
35. Kaakinen, J. W., Jorden, R. M., Lawasani, M. H., West, R. E., *Environ. Sci. Technol.* (1975) **9**, 862.
36. American Public Health Association, "Standard Methods for the Examination of Water and Wastewater," 13th ed., p. 874, Washington, 1971.
37. Symposium on the Environmental Aspects of Shale Oil Production and Processing, Am. Inst. Chem. Eng., Denver, Aug. 28–31, 1977.
38. Kolthoff, I. M., Sandell, E. B., Meehan, E. J., Bruckenstein, S., "Quantitative Chemical Analysis," 4th ed., p. 399, Macmillan, New York, 1969.
39. Davison, R. L., Natusch, D. F. S., Wallace, J. R., Evans, C. A., Jr., *Environ. Sci. Technol.* (1974) **8**, 1107.

RECEIVED August 5, 1977. This study is part of the Environmental Trace Substances Research Program of Colorado.

Aspects of Chromatographic Analysis of Oil Shale and Shale Oil

P. C. UDEN, A. CARPENTER, JR., F. P. DiSANZO, H. F. HACKETT, and S. SIGGIA

Department of Chemistry, Trinity College, Hartford, CT 06033
Amherst, MA 01003

D. E. HENDERSON

Department of Chemistry, Trinity College, Hartford, CT 06033

The various modes of high-resolution gas chromatography (HRGC) are assessed for different fractions of shale oil. Wall-coated open tubular (WCOT), support-coated open tubular (SCOT), and porous layer open tubular (PLOT) columns are compared for an alkane–alkene fraction, the WCOT giving greatest resolution, but the PLOT column having the most sample capacity and the greater compatibility with post-column identification techniques. A range of PLOT columns give good resolution for the tar-acid fraction of shale oil, free fatty acid phase (FFAP) giving the best performance. Identification of both hydrocarbon and phenolic components is accomplished by interfaced mass spectrometry and vapor-phase IR spectroscopy. Shale oil bases are separated also and identification attempted by vapor-phase IR. Pyrolysis GC (py-GC) of oil shale is accomplished using an interfaced thermal analysis system and the effect of on-line subtractive reactors is discussed. High-pressure liquid chromatography (HPLC) using subtractive reactors is accomplished also.

Until the recent years of renewed concern for alternative fuel oil and petrochemical sources, interest in the mineral shales which contain the earth's most abundant organic matter, kerogen, has been sporadic. Both in the United States and throughout the world there are vast shale reserves, but the technology to exploit them and extract the oil from kerogen by heating and refining the volatiles produced has been slow to develop, primarily owing to economic and environmental factors. Oil shale can supply a viable route to supplement the world's diminishing petroleum reserves provided that the various governmental, commercial, and other interests all rapidly develop needed technology. The part to be played by analytical chemistry is self-evident. A chemical resource which is in many respects considerably more complex than petroleum requires all the techniques of modern analysis for its characterization. The needs were made clear at a National Science Foundation-sponsored workshop (1) which considered the overall problem of analytical chemistry pertaining to oil shale and shale oil. The acceleration in the general level of interest in the field has been evidenced in two recent publications (2, 3) which cover widely the science and technology of oil shale, tar sands, and related fuel sources.

Analytical separation techniques are central to the characterization of a process as complex as the production of shale oil from oil shale and to the detailed elucidation of the nature of shale oil, spent shale, and other chemical features of shale technology. While classical separation methods have been applied to these analyses, modern high resolution chromatographic separation procedures have had only limited application to the present time, particularly HRGC and HPLC techniques. Most GC used has involved the separation of the oil, first by boiling range through distillation, and then by extraction methods to obtain a series of functional group fractions for analysis. The inability of direct GC separation to accommodate the wide chemical range of components present in shale oils has made this extensive work-up necessary.

The need for both high resolution separations and chemical class determinations is evident in order to obtain a complete characterization of a specific oil. However, chemical class analysis alone may be adequate to answer the question of the quality of shale oil obtainable from a given shale. One of the present research goals is the development of direct on-line analytical procedures for the pyrolysis products from milligram samples of shale. Py-GC with selective subtractive pre-columns between the pyrolysis reactor, and the gas chromatographic column are used. The major advantages of this method lie in the minimization of sample handling, the speed of the analysis which results from the use of small shale samples, and the fact that no bulk condensation of the pyrolysis products is necessary. Although vapor phase subtractive techniques are less selective than liquid phase methods, good analytical results have been obtained using HRGC separations. In contrast, the reduction of chromatographic resolution has made possible the quantitation of classes of compounds by comparison of the evolved gas envelope with and without the use of selective subtractive reactors. In a parallel fashion, liquid chromatographic methods are being developed to explore the use of selective resins and adsorbents for quantitation of chemical class types in the liquid phase.

Also important is the use of the modern chromatographic methods to separate and identify individual components in commercial crude shale oil. The use of efficient gas chromatographic columns coupled with ancillary techniques such as mass spectrometry and vapor-phase IR spectroscopy allows the identification of individual shale oil components. A principal part of this study is the comparison of different types of open tubular columns for the separation of the alkane–alkene fraction of shale oil; WCOT, SCOT, and PLOT columns are examined. Gas chromatographic separation of shale oil acids and bases also is performed allowing the identification of these components. The potential utility of subtractive pre-columns in HPLC analysis is illustrated also.

Instrumentation and Experimental Methods

Pyrolysis studies of oil shale, gas phase subtractive investigations of evolved volatiles, and open tubular gas chromatography—vapor-phase IR spectroscopy of shale oil components were carried out in the interfaced vapor-phase thermal analysis laboratory which has been described in detail elsewhere (4, 5). A general block diagram of this system is shown in Figure 1.

The MP–3 Thermal Chromatograph (Spex Industries, Inc., Metuchen, NJ) facilitates the slow (up to 40°C/min) controlled atmosphere pyrolysis of solid oil shale. The volatile compounds produced are monitored by both flame ionization and thermal conductivity detectors and after trapping, the whole or portions of the evolved organic profile may be further subjected to GC and other analytical procedures. In contrast, the CDS 100 pyroprobe and associated CDS 820 reaction system (Chemi-

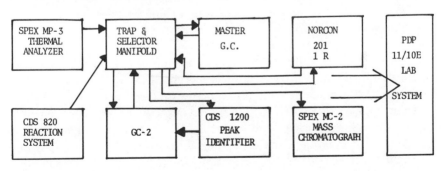

Figure 1. System block diagram of the interfaced peak identification system
(4)

cal Data Systems, Inc., Oxford, PA) enables the very rapid pyrolysis of the shale to be carried out, samples being heated at rates of up to 20,000°C/sec. By means of the central high-temperature trap manifold, evolved gas samples from either source may be directed to a master gas chromatograph (Varian 2760 equipped with both flame ionization and thermal conductivity detection) for appropriate separation prior to vapor-phase IR spectroscopy with the interfaced Norcon 201 spectrometer (Norcon Instruments, So. Norwalk, CT). Alternatively, vapor samples may be passed to a CDS 1200 instrument for elemental or functional group analysis or to a MC–2 Mass Chromatograph (Spex Industries, Inc.) for molecular weight determination.

The Norcon 201 instrument allows measurement of the IR spectra of eluent GC peaks. Its rapid scan capability (6 sec for range 2.5–15 μm) permits ready analysis within the time frame of GC in a continuous analysis mode. The CDS 1200 is a catalytic thermal degradation system which allows the determination of the elemental composition of gaseous-evolved components. The organic compound is converted to CO_2, N_2, CO, and H_2O and with the appropriate GC separation, these gases are quantitated and an elemental composition obtained for carbon, nitrogen, oxygen, and hydrogen, respectively. The MC–2 Mass Chromatograph allows the measurement of the molecular weight of the separated evolved components from the relative responses of the GC peak in CO_2 and Freon 115 carrier gases with dual gas density balance detectors. All data handling was carried out with an interfaced PDP11–10E minicomputer (Digital Equipment Corp.) which performed such functions as measurement of chromatographic retention times and peak areas, background subtraction, filtering and amplification of IR spectra, and calculation of unknown molecular weights from MC–2 data.

Materials for selective vapor-phase trapping have been considered by various workers (6, 7). Traps used in the present study included molecular sieve 5A for subtraction of straight chain hydrocarbons, sodium bisulfate and phosphoric acid for subtraction of shale oil bases, and alumina for subtraction of acidic components.

The oil shale samples and the crude shale oil used in this study were obtained from TOSCO (The Oil Shale Corporation), Golden, Colorado.

Quantitation of total shale oil bases by subtractive HPLC was carried out on a Varian 4100 instrument having UV detection at 254 nm, using a pre-column packed with Dowex 50WX8 cation exchange resin.

A range of open tubular GC columns were used. A 150 ft × 0.01 in. i.d. stainless steel wall-coated open tubular (WCOT) column (Perkin Elmer Corp., Norwalk, CT), a 50 ft × 0.02 in. i.d. support coated open tubular (SCOT) column (Perkin Elmer), and 33 ft and 100 ft × 0.03 in. i.d. porous layer open tubular (PLOT) columns. The latter were prepared in the laboratory according to the method of Nikelly (8) using stainless steel tubing (Handy & Harman Tube Co., Norristown, PA). An injection stream splitter (Perkin Elmer) was required for WCOT and SCOT columns, but direct injection was made onto the PLOT columns because they exhibit greater sample capacity.

Further chromatographic peak identification also was carried out by interfaced GC-mass spectrometry (MS) using a Hitachi–Perkin Elmer RMU6L mass spectrometer interfaced through a jet separator (Scientific Glass Engineering Pty., Melbourne, Australia).

Results and Discussion

Gas Chromatographic Separations. The aim of this chapter is to indicate where some of the more recently introduced chromatographic techniques can find application in oil shale and shale oil analysis. In this regard a study was made of the various available types of open tubular GC columns available for high resolution separations of shale oil. Open tubular GC has been used extensively for hydrocarbon analysis (9, 10) and its use for high-boiling fractions of crude oil has been illustrated for aromatic constituents boiling up to 435°C (11). A 30-ft OV 101 WCOT column has been used for heavy petroleum fractions up to carbon number 58 (12). Petroleum oil fingerprinting has been accomplished (13) and hydrocarbon oil spill characterization achieved successfully (14). Coal-derived fluids have been characterized (15) (the separation problems being more complex because of their greater content of nitrogen- and oxygen-containing compounds). Gallegos (16, 17) has used open tubular GC-MC to study functionality in methanol extracts of oil shale utilizing SCOT and WCOT columns for sterane and terpane characterization. He also used a WCOT Dexsil 300 GC column in the GC–MS mode (18).

The nonpolar fraction of shale oil was chosen to assess the relative utility of WCOT, SCOT, and PLOT columns (the latter designation referring to columns packed in the laboratory by the dynamic coating method (8)). This fraction comprises primarily saturated and unsaturated hydrocarbons, and samples were obtained by Florisil column separation (19) and elution with heptane. This same fraction may be resolved also from the oil shale pyrolyzate in the vapor phase by an on-line subtractive sodium bisulfate reactor. The different types of open tubular columns were assessed in terms of resolution, sample size capacity, upper temperature limits, and compatibility with pre-column reactors and post-column vapor-phase identification techniques. A further step in sample preparation was carried out to provide a limited molecular weight range sample to best assess the WCOT and SCOT columns; the oil fraction was subjected to preparative GC on a 10% SE30 silicone column and the fraction eluting up to 250°C was trapped in a glass loop cooled with liquid nitrogen and subsequently was washed from the loop with heptane. Portions of the chromatograms obtained with OV–1 WCOT and SCOT

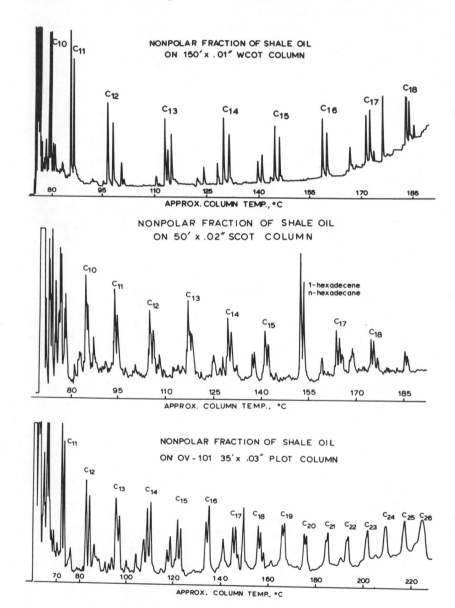

Figure 2. GC comparison of WCOT, SCOT, and PLOT columns for separation of a shale oil nonpolar fraction

columns and an OV–101 PLOT column are shown in Figure 2. It is clear that the WCOT column produces the best resolution of the sample components as is predicted although it must be noted that it was the longest column used. Separate carbon number alkane–alkene groups are well resolved, but for this column there is evidence of drift on programming which makes chromatographic difficulties above C–18. The SCOT column shows greatly reduced resolution as evidenced for the spiked hexadecane–hexadecene pair. Early band broadening caused by the necessary use of the injection splitter is more in evidence than for the shorter PLOT column where direct injection was possible. Overall resolution for the latter column is close to that of the SCOT column and it exhibited a considerably higher temperature limit than the others. Its sample capacity is considerably greater making its applicability for quantitative study and interfaced identification techniques superior. In general, WCOT and PLOT columns appear to form a very useful complementary pair for resolution, quantitation, and identification studies.

Since PLOT columns, although applied in other areas (8, 19), have received little attention in the shale oil field, a number of applications in different areas will be considered. Their ready laboratory preparation makes the evaluation of novel stationary phases straightforward and their sample level compatibility with such interfaced techniques as vapor-phase IR makes them very attractive.

A chromatogram of crude shale oil with no separative work-up on the 35-ft OV–101 PLOT column is shown in Figure 3. Comparison with the PLOT chromatogram in Figure 2 shows that the resolution of the alkane–alkene pairs is virtually unaffected by the presence of the other oil components.

Despite the usual preference in GC for the application of more polar stationary phases for the resolution of polar species, the former may be applied usefully for nonpolar separations. Figure 4 illustrates the resolution of the alkane–alkene pairs of the hydrocarbon fraction of shale oil (sample as in Figure 2), on a FFAP PLOT column (100 ft). FFAP is Carbowax 20M terminated with terephthalic acid residues. This column selectively retains olefins and enables base-line resolution of the alkane–alkene pairs. The only disadvantage of this phase is its temperature limitation of ca. 250°C. Identification of the numbered peaks in Figure 4 was achieved by interfaced MS (components are listed in Table 1).

On considering the optimal chromatographic conditions for analysis of the more polar constituents of shale oil it becomes apparent that both packed columns and PLOT columns have useful features. An example involves the acid-component fraction of shale oil obtained by sodium hydroxide extraction, removal of nonpolar species with benzene, methyl-

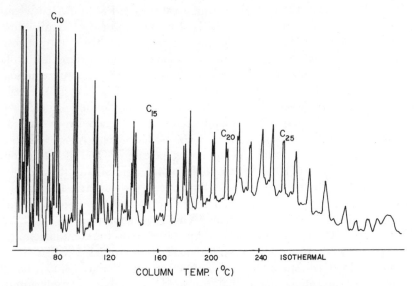

Figure 3.   Gas chromatogram of Green River shale oil on a 35 ft × .03 in. i.d. OV 101 PLOT column. Substrate Chromosorb R6470–1.

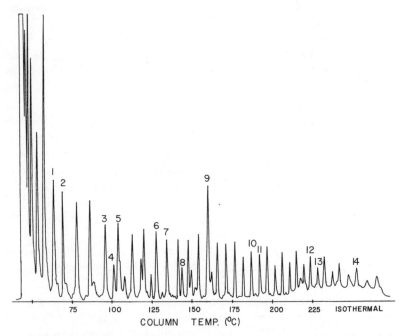

Figure 4.   Gas chromatogram of nonpolar fraction of TOSCO II (Green River) shale oil on a 100 ft × .03 in. i.d. FFAP PLOT column. Substrate Chromosorb R6470–1 temperature program 40°–225°C at 4°C/min.

Table I. Identification of Hydrocarbon Peaks in Figure 4
(numbered) by Interfaced Mass Spectrometry

Peak Number	Identity	Peak Number	Identity
1	n-undecane	8	$C_{18}H_{38}$ branched chain alkane
2	1-undecene	9	1-heptadecene + C_{19} branched olefin
3	n-tridecane	10	n-eicosane
4	$C_{14}H_{28}$ branched alkene	11	1-eicosene
5	1-tridecene	12	n-tetracosane
6	n-pentadecane	13	1-tetracosene
7	1-pentadecene	14	n-heptacosane

ene chloride, and diethyl ether, and back extraction with diethyl ether after neutralization with hydrochloric acid.

Figure 5 shows the separation of this fraction on a packed 6-ft Dexsil 410 GC column. (Dexsil 410 GC is a cyanosilicone–carborane phase of

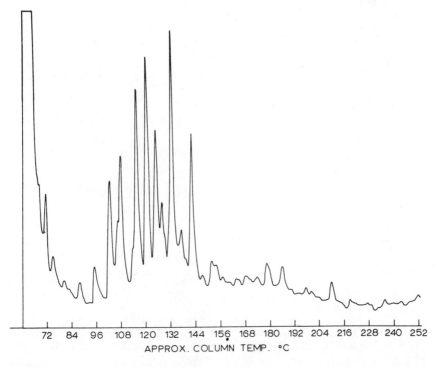

APPROX. COLUMN TEMP. °C

Figure 5. Gas chromatogram of base-soluble species from Tosco II shale oil on a 6 ft × 1/8 in. o.d. packed column. 3% Dexsil 410 GC on Chromosorb 750.

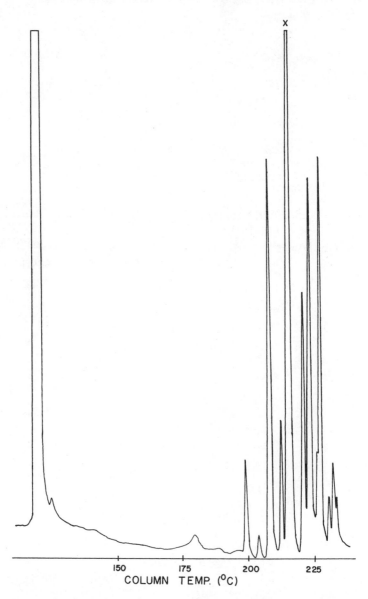

Figure 6.   Gas chromatogram of shale oil acids extracted from Tosco II shale oil on a 100 ft × .03 in. i.d. FFAP PLOT column. Substrate Chromosorb R6470–1 temperature program 90°–225°C at 6°C/min.

moderate polarity and temperature limits in excess of 350°C). The peak resolution is good for these acidic compounds; by comparison virtually no resolution is obtained on nonpolar methyl or phenyl silicone phases. The PLOT FFAP column noted previously also was used for the acids analysis and the separation obtained is shown in Figure 6. It is evident that resolution is improved considerably with base-line separation between most components although elution temperatures are considerably higher than for the Dexsil column. This limits the utility of FFAP for high-boiling shale acids, but does not interfere with the analysis of most of those present.

This PLOT separation allowed vapor-phase IR and mass spectral determinations to be made. Figure 7 shows a typical IR spectrum obtained from Peak X in the above FFAP chromatogram. The spectrum corresponds to a 6-sec scan with computer background correction and normalization. The spectral features correspond well with an alkyl phenolic structure and a comparison spectrum is shown in Figure 8 for 2,4-dimethylphenol eluted from the FFAP column under identical condi-

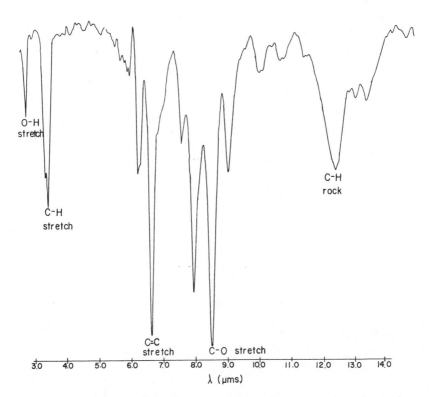

Figure 7. Vapor-phase IR spectrum of acidic component Peak X (Figure 6). Helium flow rate 10.5 mL/min, 6-sec scan.

tions; this compound having the same retention time as the peak in question. The features of the two spectra are closely similar, variations in relative intensity being attributable to concentration changes as the band moves through the cell. Other small dissimilarities may result from the presence of minor unresolved components. In confirmation, mass spectral data for a low-energy spectrum of Peak X (9.5 eV) shows a molecular ion at 122 confirming a molecular formula of $C_8H_{10}O$. Further detailed studies on these tar-acid constituents are in progress and will be reported separately (20).

The gas chromatographic separation of the shale oil bases is a very complex problem because of the great number of components of this type present. An initial acid extraction followed by an alumina column fractionation is necessary before GC. The FFAP PLOT column again proved optimal for this pyridine–quinoline base fraction as shown in Figure 9. A vapor-phase IR spectrum of Peak A in this chromatogram is shown in Figure 10, displaying the typical characteristics of an alkyl-substituted pyridine. The mass spectral fragmentation pattern also con-

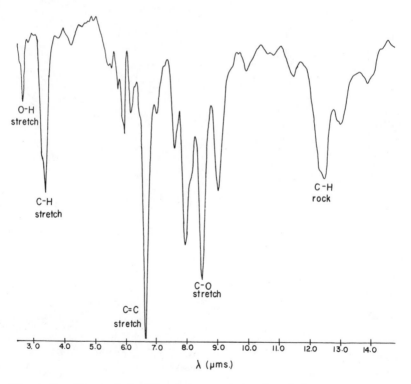

Figure 8. Vapor-phase IR spectrum of 2,4-dimethylphenol. Helium flow rate 10.5 mL/min, 6-sec scan.

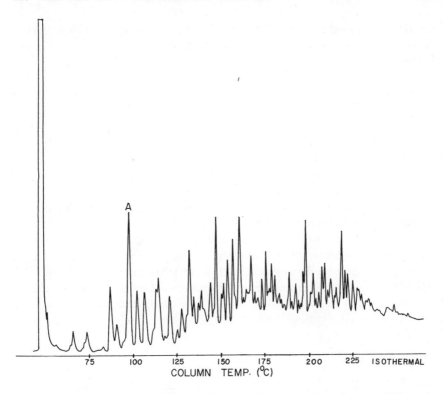

Figure 9. Gas chromatogram of shale oil bases after acid extraction and alumina column fractionation. 100 ft × .03 in. i.d. FFAP PLOT column. Substrate Chromosorb R6470–1. Temperature program 50°–225°C at 6°C/ min.

firms this assignment. From the GC–IR and GC–MS data this component was identified as 2,4-dimethyl-6-ethylpyridine.

These chromatographic examples demonstrate the utility of PLOT columns for the resolution of the various chemical constituent fractions of shale oil and their compatibility with interfaced IR and mass spectral peak identification. The major advantage of the higher column peak capacity of the PLOT columns makes the latter measurements more feasible particularly for the minor components in the chromatogram.

Oil Shale Pyrolysis-Gas Chromatography. Another important aspect of gas chromatography as applied to oil shale characterization is the ability to carry out separations directly on volatiles evolved as samples are heated under controlled conditions. The method for heating small samples of oil shale in the laboratory clearly differs considerably from actual larger scale retorting procedures. In the latter case, secondary reactions of organic pyrolysis product certainly occur and they are subject to contact with water and air prior to work-up and analysis. However, a small scale laboratory procedure may produce at least partial simulation of retorting procedures under conditions where organics are largely swept away from the reactor zone as they are produced. It yet remains to be seen how good a model this procedure is for retorting and whether the analytical techniques produce a viable indication of the final products produced in a retort. Preliminary results from the vapor-phase analysis system are encouraging and there seems to be fair general correspondence between overall py-GC profiles and commercial shale oils. A more detailed study of the degree of qualitative and quantitative correlation for specific chemical classes is underway presently (20).

The thermal chromatograph is the instrument of choice for the heating of small (mg) shale samples. These are heated under controlled conditions of temperature increase under a chosen atmosphere, and volatiles produced are carried directly through thermal conductivity and/or flame ionization detectors. After trapping of all or part of the evolved species, GC may be carried out. A typical pyrolysis profile of the total volatiles produced with respect to temperature for both types of detection is shown in Figure 11. The characteristic twin-humped curve is seen with thermal conductivity detection. The latter hump, which is absent

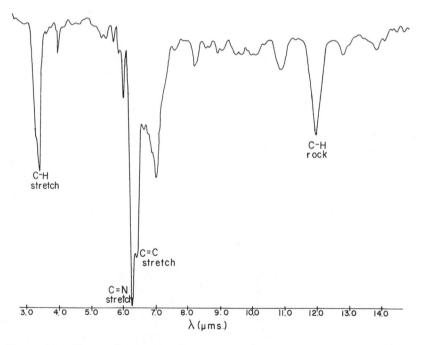

Figure 10.   Vapor-phase infrared spectrum of basic component Peak A (Figure 9). Component identified as 2,4-dimethyl-6-ethylpyridine. Helium flow rate 10.5 mL/min, 6-sec scan.

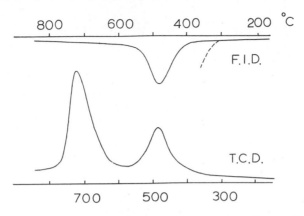

Figure 11. MP–3 thermogram of Green River oil shale. 6 mg of 60/80 mesh sample heated at 10°C/min in nitrogen (40 mL/min).

in the flame ionization detector trace, corresponds to carbon dioxide evolution. Thus far, GC analysis with this system has been carried out primarily with packed columns. A feature of the instrumental system is the ability to pass the evolved volatiles envelope through subtractive loops to remove specific species selectively. This may be done either before the first chromatographic analysis or at any time subsequently, enabling direct comparisons for the same sample to be made. Figure 12 shows the results of such an experiment. The full chromatogram is that obtained after the total organic volatiles produced up to 550°C (Figure 11) were passed through a subtractive reactor consisting of sodium bisulfate coated on Chromosorb W in a 1:1 ratio. The dashed line represents the position of the base line of the unresolved envelope chromatographed directly before passage through the reactor; the same resolved peaks as in the reacted case are superimposed on the unresolved envelope. The difference between the two traces represents the species removed by the acidic reactor and these presumably correspond to the tar bases. It is interesting to note that a comparison carried out between the alkane–alkene fraction of shale oil obtained as noted previously and a sample of oil passed through a bisulfate reactor in a similar fashion gave almost identical PLOT chromatograms indicating that the bases had been removed (as in Figure 2). Small discrepancies in detail were probably caused by small amounts of tar acids and other non-basic components still present.

Another subtractive pre-column which shows potential utility is molecular sieve 5A. This material is known to selectively subtract straight chain organic species, although compounds other than alkanes are removed. Table II shows results obtained at different final pyrolysis temperatures for oil shale samples pyrolyzed in the thermal chromatograph.

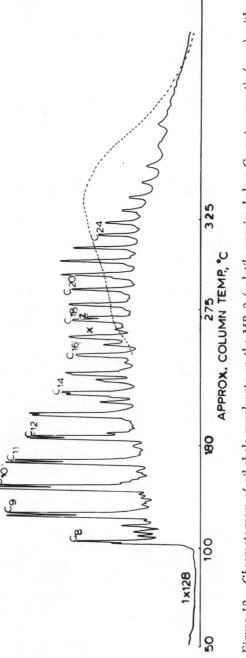

Figure 12. Chromatogram of oil shale pyrolyzate on the MP–3 (volatile species below C_8 not present): (– – –) without NaHSO_4 reactor; (———) with NaHSO_4 reactor; 6 ft × 1/8 in. o.d. SE30

Table II. Shale Oil Composition by Vapor-Phase Methods

Temperature (°C)	% Weight Loss	% H₂O	% Oil	% Straight Chain
510°	15.6 ± 1.0	8.97	91.0	43.1
540°	16.2 ± 1.7	9.26	90.7	43.4
610°	17.3 ± 0.9	9.83	90.2	46.4

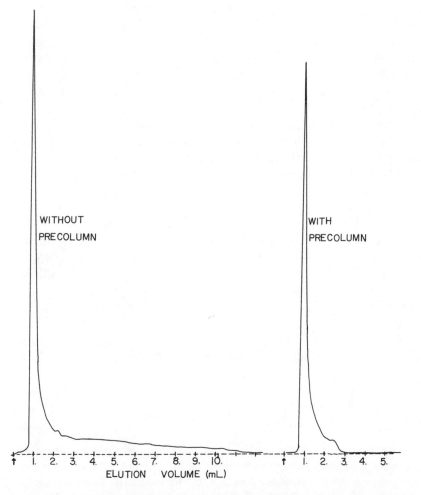

WITHOUT PRECOLUMN

WITH PRECOLUMN

ELUTION   VOLUME (mL)

Figure 13.   HPLC of total shale oil. Cation exchange precolumn subtraction of shale oil bases. Conditions to retain the most basic components only. Column, 4 mm × 25 cm Partisil 10. Mobile phase, 60/30/10, C_6H_{14}/CH_2Cl_2/THF. Flow rate .50 mL/min. Pre-column 1/8 in. × 4 in. Dowex 50 W × 8 Resin pre-activated.

The percentage of water produced was determined independently on a 6-ft Chromosorb 103 (Johns Manville Corporation) column using a thermal conductivity detector. Thermal conductivity was used also for the organic quantitation, since its response is affected little by carbon number. Gas chromatograms of the total evolved organics were obtained before and after passage through the molecular sieve pre-column, on a short packed SE 30 column. Average areas per sample weight with and without the pre-column were subtracted to determine the composition. These results, although preliminary, indicate that selective pre-columns may be useful for bulk sample type analysis in low resolution situations as well as for the specific abstraction of individual compounds HRGC.

The potential of this type of on-line reactor for chromatographic differentiation of portions of very complex mixtures is considerable. We presently are considering the problems involved in quantitatively regaining the subtractive components from such reactors for confirmatory analysis.

High-Pressure Liquid Chromatography. There appears to be little doubt that HPLC will play a prominent part in future separation of shale oil components. Liquid-phase methods have long been important in work-up procedures for analysis and their extension to high resolution areas is under extensive study. One particular example is chosen to illustrate the capabilities of HPLC; that of its combination with reactive pre-columns for quantitative estimation of oil compound types.

Figure 13 shows a total shale oil liquid chromatogram in the absence and in the presence of a cation exchange pre-column. The subtracted basic components are quantitated from the difference in areas with and without the pre-column. The subtracted basic species were determined as ca. 12.9% of the total crude shale oil using this method.

It is clear that in the future high-resolution liquid-phase methods will be used increasingly in shale oil analyses, both in the separation mode and also as illustrated here in the subtractive mode for group quantitation.

Acknowledgments

We wish to acknowledge the support of the National Science Foundation for this program through Grant CHE74 15244. We also acknowledge support for the establishment of the Vapor-Phase Thermal-Analysis Laboratory from the National Science Foundation through Research Instrument Grant GP 42542 to the University of Massachusetts. We thank M. T. Atwood of The Oil Shale Corporation (TOSCO) for samples of oil shale and shale oil.

Literature Cited

1. Siggia, S., Uden, P. C., Eds., "Analytical Chemistry Pertaining to Oil Shale and Shale Oil," Report of NSF Conference–Workshop, Washington, D.C., 1974.
2. Yen, T. F., Ed., "Science and Technology of Oil Shale, Ann Arbor Science, Ann Arbor, Michigan, 1976.
3. Yen, T. F., Ed., "Shale Oil, Tar Sands, and Related Fuel Sources," ADV. CHEM. SER. (1976) **151**.
4. Uden, P. C., Henderson, D. E., Lloyd, R. J., *J. Chromatogr.* (1976) **126**, 225.
5. Uden, P. C., Henderson, D. E., Lloyd, R. J., "Analytical Pyrolysis," p. 351, Elsevier, Amsterdam, 1977.
6. Ettre, L. S., McFadden, W. H., "Ancillary Techniques of Gas Chromatography," John Wiley, New York, 1969.
7. Leathard, D. A., Shurlock, B. C., "Identification Techniques in Gas Chromatography," John Wiley, New York, 1970.
8. Nikelly, J. G., *Sep. Purif. Methods* (1975) **3**, 423.
9. Bradley, M. P. T., *Anal. Chem.* (1975) **47**(5), 181R.
10. Bradley, M. P. T., *Anal. Chem.* (1977) **49**(5), 249R.
11. Stuckey, C. L., *J. Chromatogr. Sci.* (1971) **9**, 575.
12. Gouw, T. H., Whittemore, I. M., Jentoft, R. E., *Anal. Chem.* (1970) **42**, 1394.
13. Adlard, E. R., Powell, J. L., *Environ. Sci. Technol.* (1976) **10**, 284.
14. Rasmussen, D. V., *Anal. Chem.* (1976) **48**, 156.
15. Bertsch, W., Anderson, E., Holzer, G., *J. Chromatogr.* (1976) **126**, 213.
16. Gallegos, E. J., *Anal. Chem.* (1971) **43**, 1151.
17. Gallegos, E. J., *Anal. Chem.* (1973) **45**, 1399.
18. Gallegos, E. J., *Anal. Chem.* (1975) **47**, 1524.
19. Dineen, G. U., *Anal. Chem.* (1955) **27**, 185.
20. Siggia, S., Uden, P. C., Carpenter, A., Jr., Hackett, H. F., to be published.

RECEIVED October 17, 1977.

16

Olefin Analysis in Shale Oils

L. P. JACKSON, C. S. ALLBRIGHT, and R. E. POULSON

Energy Research and Development Administration, Laramie Energy Research Center, Laramie, WY 82071

The classical silica gel chromatographic method for determination of percent olefins in shale oils was studied and found wanting, mainly because of cross-contamination from high levels of olefins and heteroatom-containing compounds. This paper describes a new hydroboration/oxidation procedure for olefins, and reports its use in hydrocarbon-type analysis of both whole shale oils and distillate fractions. Percent composition values for three whole oils ranged as follows: saturates, 13–26; olefins, 16–20; aromatics, 5–14; polar compounds, 41–62. A discussion of IR analysis for relative amounts of specific olefin types such as terminal, internal trans, and methylene structures is included.

The analysis of shale oils presents a series of problems to the chemist which, until recently, have not been encountered in the analysis of total petroleum products. Two of the most serious problems among these have been the high levels of olefins and heteroatom-containing compounds. Increased industrial production of materials from coking operations and preliminary excursions into synfuels have brought these problems to the attention of petroleum analysts and have made the development and dissemination of new analytical techniques necessary.

Work in the area of hydrocarbon-type analysis, i.e. saturate, olefin, and aromatic, has been conducted at the Laramie Energy Research Center (LERC) for over two decades. During this time, a significant portion of the effort has been directed towards the quantification and characterization of the various types of olefins found in Green River Formation shale oils and their distillate fractions (1–9).

The olefin concentrate was prepared by some variation of column chromatography on silica gel using pentane as an eluent for the oil or isopropyl alcohol to displace the oil. In all cases, rather diffuse boundaries between the respective hydrocarbon types were obtained with some cross-contamination. The difficulty with cross-contamination increased markedly upon going to successively higher boiling distillate cuts (7), and total oils were beyond the capability of the method. Owing to these deficiencies of the silica gel method we have begun developing a new method for olefin analysis. The basis of the method is the facile addition of diborane and its organoborane analogs across double bonds followed by the ready cleavage of the addition products to yield alcohol derivatives of the olefins. These derivatives are more amenable to quantification, characterization, and chromatographic manipulation than the parent olefins.

This chapter presents a close look at the standard silica gel analysis of a shale oil naphtha with an evaluation of its effectiveness, a description of the new method now used to quantify olefins in shale oil products, a summary of results of the hydrocarbon-type analysis using the new method for a series of three related shale oils, and a discussion of the information on olefin-type compounds which can be revealed by IR examination of whole shale oils. The paper concludes with a brief discussion of additional applications of the hydroboration of olefins to problems of interest to the petroleum analyst.

Experimental

Hydrocarbon Types in Naphtha. A shale oil neutral naphtha (ibp to 220°C) was prepared by successive extractions of the naphtha sample with 10% sulfuric acid to remove the tar acids and tar bases, respectively. A 12-g portion of the neutral naphtha was charged to a column (8-mm i.d. $\times$ 200-mm long, water-jacketed) containing 60 g of activated (150°–200°C, 3 hr) code 922 silica gel (thru 200 mesh). Pressure ($\sim$ 3 lb of N_2), was applied to the top of the column until almost all of the sample had entered the bed. The pressure was released slowly and the sides of the column washed down with a small portion of isopropyl alcohol. When all of this had entered the column, a few grams of fresh gel were added to the top of the column followed by 100 mL of isopropyl alcohol. The column was developed under pressure (8–14 lb, N_2) at a constant rate. The progress of the displacement was followed by obtaining the refractive index (R.I.) on each 0.14-mL fraction. Separation into the three hydrocarbon types was based on the inflection points in the curve obtained by plotting R.I. vs. fraction number. The average values of three duplicate runs were: 29% saturates, 47% olefins, 21% aromatics, and 3% polars.

Saturates. Acid absorptions were performed on higher-boiling distillate fractions by a GC-internal standard method as described elsewhere (9). Briefly, the values of saturate content were determined by GC comparison of sample-to-internal standard ratios before and after absorption. Acid absorption on whole oils was performed on 10 mL of a 20% solution in cyclohexane as described in ASTM D 1019–68 (10). The saturates were recovered from the solvent and weighed directly to determine percent composition.

Olefins; Aromatics. The hydroboration and analysis of shale oil distillate fractions have been described previously (9). Whole shale oils were prepared for determination of olefin and aromatic content by the removal of the polar material on Florisil (30/60 mesh, used as received) by elution with cyclohexane. The Florisil-to-oil weight ratio was 20/1 and the elution was carried out for 24 hr with a solvent recycle chromatography column.

The neutral oil obtained from the Florisil chromatography was assumed to have an average molecular weight of 226, corresponding to $C_{16}H_{34}$. Enough of the borane–methyl sulfide complex (Aldrich Chemical Company) was used to react with the oil assuming 100% olefinic content. In a typical determination, 1.34 g (6.0 $\times$ 10^{-3} mol) of neutral oil was dissolved in 20 mL of tetrahydrofuran (THF) that had been freshly distilled from LiAlH$_4$; this solution was added to a dry reaction assembly consisting of a 100-mL, three-necked flask equipped with a magnetic stirring bar, reflux condenser, thermometer, and a dry N$_2$ purge which was introduced into the flask through a rubber septum affixed to one of the three necks. The borane–methyl sulfide complex (0.20 mL, 2 $\times$ 10^{-3} mol) was dissolved in 10 mL of dried THF and added slowly to the reaction flask through the septum. After the addition was complete an additional 10 mL of dry solvent was added and the solution was brought to reflux with a heating bath. The solution was stirred at reflux for 17 hr at which time the heat was removed and the solution allowed to return to room temperature. Then 2 mL of methyl alcohol was added very slowly to the reaction mixture (caution: vigorous reaction) to destroy excess reagent. This was followed by 0.67-mL 3N sodium hydroxide and then 0.7 mL of 30% hydrogen peroxide (caution: vigorous gas evolution). The nitrogen purge was discontinued and the solution refluxed for 1 hr. After the solution had cooled, the reaction mixture was added to a separatory funnel, diluted with 75 mL of cyclohexane, and washed 3 $\times$ 100 mL with water followed by a single wash with 50 mL of saturated NaCl solution. The resulting cyclohexane solution was dried over MgSO$_4$, filtered, and the solvent was removed on a rotary evaporator with a slight vacuum. The yield was 1.42 g of a mixture of saturates, aromatics, and the alcohol derivatives of the olefins.

The saturate and aromatic hydrocarbons were separated from the alcohols by elution chromatography on alumina deactivated with 3.8% water in a ratio of 5 g of alumina to 1 g of oil, by successive elutions with (1) 30 mL of cyclohexane and (2) 50 mL of 2% benzene in cyclohexane. The alcohols were recovered for future analysis by washing the column with 100 mL of ethyl ether. The solvent was removed from the hydrocarbon fraction and the yield of saturates plus aromatics calculated back to the neutral oil. The percentage of olefins was determined by difference. The percentage of aromatic material was determined as the difference between the saturates and aromatics from the hydroboration and the saturates from acid absorption.

Boiling Range Distribution. The distribution of boiling range was obtained by a simulated distillation procedure, using an automated GC equipped with a flame ionization detector (FID) using automatic data acquisition and reduction units. The column was 3/16 in. $\times$ 10 ft SS-packed with 3% OV-1 on high performance Chromosorb W, acid washed, with a 70 mL min^{-1} flow of helium.

Using the normal alkanes C_5 through C_{42} as a standard, the column was calibrated to give a boiling point vs. time relationship and the automated data readout reported the results of each simulated distillation as percent oil off for each 100°F increment through 1000°F. The values for residuum—material boiling above 1000°F—were obtained by comparing the total response of an external standard having a 450°–800°F boiling range with the total response of the oil, the difference being residuum. Both the external standard and the oil were corrected for background by running blanks and making suitable corrections. A standard was run for each determination and the oils were run in duplicate. Each oil was corrected for solvent content as necessary, and experience showed that the results of duplicate simulated distillations agreed within ~ 0.2% absolute for any given 100°F fraction.

IR Analysis. IR absorption spectra were determined on neat samples of shale oils to furnish data for estimates of various olefinic types of compounds. Samples were run on a high-resolution, double-beam grating spectrophotometer at 0.1-mm path length between KBr plates. Quantitative measurements were made using the cut and weigh method with baselines drawn from point to point of minimum absorption.

Results and Discussion

The experimental procedure followed in the silica gel separation of the shale oil naphtha closely follows ASTM D 2003–64 for the determination of saturates in high-olefinic petroleum naphthas (*11*). The significant differences are that the sample must be prepared by the removal of tar acids and tar bases prior to silica gel separation, and the R.I. of the material is used to determine the break points in the composition as in the method of Saier (*12*). It was not designed for the preparation of olefinic concentrates. The ASTM D 2549–68 method is designed to provide a saturate and aromatic concentrate from higher boiling petroleum fractions (450°–1000°F). The method has no value in the determination of olefins because it combines the saturates and olefins in the same fraction and does not differentiate aromatic material containing double bonds in side chains from those containing none.

The three fractions resulting from the silica gel chromatography of the shale oil neutral naphtha were examined in detail to provide information regarding the degree of cross-contamination of the various fractions. The saturate's fraction was examined by IR, UV, and proton NMR and showed no aromatic or olefinic contamination, but when subjected to the acid absorption procedure it gave a 3% loss. The lack of any UV, IR, or NMR absorptions owing to aromatic or olefinic materials accompanied by the shrinkage upon acid absorption indicates that the nonsaturate material is probably tetra-substituted olefins. The low concentration, however, may preclude the detection of the lone proton on tri-substituted olefins by proton NMR or IR, therefore they cannot be ruled out as possible contributors to the nonsaturate material.

The olefinic fraction gave no indications in the UV or proton NMR analysis of any aromatic contamination and it would be impossible to detect saturates in this fraction by the two techniques. Acid absorption analysis of this fraction gave 95% shrinkage, indicating a 5% saturate content. Values of average carbon number and substitution patterns calculated for each fraction by mass spectroscopy (MS) and NMR indicated that the aromatic fraction was contaminated by ∼ 12% olefins. These may be olefinic side chains or individual olefins or perhaps dienes which may account for their being eluted later than most olefins. The IR work reported later gives preliminary indications of significant diene concentrations in shale oils. When all of these results are considered together it is found that ∼ 7% of all the olefins in the neutral naphtha are not present in the olefinic concentrate and are distributed about equally in the saturate and aromatic concentrates.

Another study from the LERC laboratories illustrates the increasing inadequacy of this type of separation as the molecular weight of the oil is increased (7). On a high-resolution liquid chromatography (LC) column (10,000 theoretical plates) which gave a wide separation of n-hexadecane from n-hexadecene-1, a shale oil saturate–olefin concentrate from a light distillate (400°–600°F) showed only an inflection point for the saturate, hindered-olefin break. A sample of heavy gas oil (800°–1000°F) saturate–olefin concentrate did not show even that. We must conclude from these data that the silica gel separation probably separates only mono-unhindered olefins from the saturates and aromatics with any degree of effectiveness.

We have found that chemical modification of the olefins with any of a variety of borane reagents followed by oxidation allows for their quantitative removal from shale oils by elution chromatography on deactivated alumina. The method is so thorough that aromatic olefins such as indene are removed completely. The reagents which have been used successfully are borane in THF, borane–methylsulfide complex (BMS), and 9-borabicyclo [3.3.1.]nonane-(9-BBN), all of which are commercially available. The use of borane in THF has been discontinued in our laboratory because BMS gives the same result, is more stable, and is easier to handle.

As with all shale oils, the polar material first must be removed from the oil prior to hydrocarbon-type analysis. The procedure shown in Figure 1 was applied several years ago to shale oil distillates in the boiling ranges 200°–325°C and 325°–500°C (9). The technique made use of added internal standards and GC comparison of sample–internal

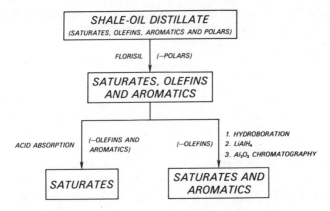

Figure 1. Separation scheme for the determination of hydrocarbon types in shale oils

standard ratios before and after treatment of the oils. In general, the values for each determination were reproducible to less than 1% of the absolute value of each constituent determined directly. The results of this study are presented in Table I. Samples I, II, and III are light distillate fractions and Sample IV represents a heavy distillate.

This same technique was applied to three whole shale oil products from our controlled-state retort which has been described previously (*13*). The polars were removed on Florisil, and the hydroboration/oxidation and acid absorption operations were carried out on samples from 1 to 2 g which were large enough to allow for gravimetric determination of material lost on treatment. The methods usually give results which agree within 2% for each material determined directly. Results of analysis of the three oils are presented in Table II. Oils A, B, and C were produced in the retort, each at a different heating rate: 0.04°F, 2°F, and 20°F min⁻¹, respectively.

Table I. Hydrocarbon-Type Analysis of Shale Oil Distillates

	% Hydrocarbon Types		
Sample	*Saturate[a]*	*Olefinic[b]*	*Aromatic[c]*
I[d]	72.9	8.1	19.0
II[d]	7.9	15.0	77.1
III[d]	45.9	27.5	26.6
IV[e]	42.8	33.2	24.0

[a] Determined by duplicate acid absorption.
[b] Determined by triplicate hydroboration/oxidation.
[c] Determined by difference.
[d] Light distillate (200–325°C).
[e] Heavy distillate (325–500°C).

Table II. Composition Analysis of Whole Shale Oils from
LERC Controlled State Retort

Oil	Heating Rate, ($°F, min^{-1}$)	% Composition			
		Polars	Saturate	Olefinic	Aromatic
A	0.04	41	26	19	14
B	2	56	19	20	5
C	20	62	13	16	9

In an effort to gain more information concerning the various olefin types present in Oils A, B, and C, the IR absorption spectra were obtained on the neutral oil produced by the removal of the polar material on Florisil. The spectra were run neat in a 0.1-mm KBr cell. The areas of the absorption peaks were determined by the cut and weigh method using baselines drawn from point to point of minimum absorption. The five peaks thus measured were located at 1640, 995, 968, 912, and 890 cm^{-1}. Peak width at half height also was determined on each peak. Using the absorbance values for each peak, the yield data for % neutral oil, and % olefin in each oil, absorbance values were back-calculated for the whole oils and pure olefins. These data are given in Table III. For any further comparison to be made, it must be assumed that the average extinction coefficient of each absorbing type is the same in each of the three oils. While examination of these data show some differences between the oils, care must be taken in making any firm assertions as to their meaning at this time.

Two of the more striking anomalies are discussed in the following sections. Oil C, which was produced at a retort heating rate of 20°F/min,

Table III. Infrared Absorption (Area) of Olefinic Bands
for Olefins from Three Shale Oils

Oil	Absorption Band, cm^{-1}					% Olefin in Oil
	1640	995	968	912	890	
A[a]	53.4	25.0	135.7	120.3	32.2	
A[b]	10.1	4.7	25.6	22.7	6.1	19
B[a]	44.0	32.0	96.9	129.3	23.6	
B[b]	8.7	5.5	16.6	22.1	4.0	20
C[a]	109.3	78.3	136.7	271.2	58.3	
C[b]	17.4	12.5	21.8	43.3	9.3	16

[a] Determined on neutral oil, corrected to 100% concentration, 0.1 mm path length, KBr cell.
[b] Determined on neutral oil, corrected to concentration in total oil, 0.1 mm path length, KBr cell.

gave the lowest yield of olefins in the total product, yet it has significantly more intense absorbance in four of the five bands which were measured when compared with the other two oils. It is not likely that this change is caused by a large decrease in the average molecular weight of the olefinic material in Oil C when compared with the other oils because on close inspection of the data in Table IV, Oil A has about the same distribution of material by boiling range. The large increase in intensity in the carbon–carbon double bond stretching frequency at 1640 cm^{-1} cannot be caused by conjugation with an aryl group on another double bond because such conjugation lowers the frequency by 15 to 40 cm^{-1}.

On the surface, it might appear that more of the material in Oil C has a terminal carbon–carbon double bond of the mono-substituted ethylene type than the other oils because movement of an internal double bond to the end of the chain causes a large increase in intensity. This also would explain the increased intensity of the carbon–hydrogen out-of-plane deformation bands of the CH and CH$_2$ groups at 995 and 912 cm^{-1}, respectively. However it should be pointed out that this high concentration of terminal double bonds in Oil C does not occur at the expense of di-substituted ethylenes which absorb at 890 cm^{-1} (CH$_2$ out-of-plane deformation) because they also show a large increase in intensity when compared with Oils A and B. The concentration of trans double bonds in Oil C lies between A and B so the difference does not lie there. It is most likely that somewhere between the production of Oil B at 2°F/min and Oil C at 20°/min the temperature threshold was reached which led to the production of polyenes in increased concentration. This would account for the increased band intensities observed. If further investigation confirms this hypothesis, this increase in band intensity could serve as a semiquantitative measure of the difference in polyene content between two oils of equal molecular weight distribution.

Oils A and B, which have essentially equal olefin contents, show differences in the intensity of the various olefinic absorbances with Oil B generally having the lesser values. These differences may all be caused by the increased molecular weight of the olefinic material in B as compared with A, which is apparent upon inspection of the percent distribution of compound types by boiling range. The trans olefin types exemplified by the absorption at 965 cm^{-1} show an approximate 50% increase in intensity on going from Oil B to A as does the asymmetric di-substituted ethylene band at 890 cm^{-1}. These increases may be too great to be accounted for by molecular weight changes alone but no other reasonable alternative presents itself at this time. Some of these changes in total olefin content may be caused by cis olefins, but the bands owing to this specie are ill-defined and variable, and no reliable data can be obtained on these olefins by this technique.

Table IV.   Percent Hydrocarbon-Type Analysis
Boiling Range (°F)

Oil	To 100	−200	−300	−400	−500
Oil A	0	0.3	3.7	7.6	10.7
Saturate	0	0	0	2.3	3.3
Olefinic	0	0	1.2	3.7	3.7
Aromatic	0	0	0	0	2.0
Oil B	0	0.4	1.7	4.6	7.3
Saturate	0	0	0.9	2.3	2.8
Olefinic	0	0	0.3	1.6	2.1
Aromatic	0	0	0	0	0.2
Oil C	0	0.4	3.1	5.6	7.2
Saturate	0	0	0.3	1.5	2.0
Olefinic	0	0	0.5	2.5	3.2
Aromatic	0	0	0	0	0.4

One more point should be brought out at this time; all of the observed bands are sharp, their position is constant, and their half-band widths are approximately equal for all of the oils indicating that there is little change in the average olefin type represented by each band in the three oils. A more detailed discussion of the IR characteristics of olefins may be found in the book by Bellamy (14) and the work by McMurry and Thornton (15).

This approach was developed to be of practical use to the process chemist who is always keenly interested in how production variables affect his product mix. Using the separation sequence shown in Figure 1 and simple subtraction techniques, it is possible to gain additional information about the three hydrocarbon types and the polar materials present in the oils. For example, knowledge of the percent composition for an oil, coupled with an accurate simulated distillation of the four fractions and the original oil, can give some insight into the quality of the oil which was unavailable as long as the various compound types were intimately mixed. The data in Table IV were determined in this manner on the three oils just discussed. The data presented in the table show the distribution of the three hydrocarbon types as a function of boiling range in 100°F increments. Of a more fundamental nature, the trends in hydrocarbon types by boiling range and the various olefin types present may give valuable insight to various reaction mechanisms that are operative during the retorting process. The data presented here are being evaluated currently to learn if they reveal anything about kerogen structure and/or pyrolysis mechanisms during oil production.

Additional applications of the hydroboration reaction which we are exploiting for information about the olefin types in shale oil are:

by Boiling Range for Three Whole Shale Oils

Boiling Range (°F)

−600	−700	−800	−900	−1000	Residuum
13.8	14.6	11.1	16.8	4.5	16.9
4.5	4.0	3.9	5.0	1.6	0.9
4.4	2.5	0	3.8	0	0.3
2.3	3.6	3.1	1.8	1.0	0.8
10.9	12.8	11.0	22.4	14.0	14.9
3.5	3.1	2.8	2.9	0.7	0
3.4	3.1	1.7	5.6	2.2	0
0.3	0.8	0.9	1.7	1.3	0
9.4	10.4	8.7	18.6	13.8	22.8
2.3	1.8	1.7	2.6	0.8	0.1
3.0	2.1	0.3	3.1	0.5	0.6
1.2	1.6	1.8	1.9	1.7	0.4

(1) Addition of the elements H–OH and D–OH across double bonds in a specific manner so that exact positions of double bonds may be determined by additional analysis.

(2) Addition of D–D across double bonds to obtain the number of double bonds, aromatic rings, and alicyclic rings in each olefin by GC–MS. In addition we are applying the method to tar sands products, coal liquids, and cycle oils and coker distillates from the petroleum industry. The results of these studies will be published when they are completed.

Summary

The classical silica gel olefin analysis was evaluated with respect to shale oil products and found wanting. A new approach to olefin analysis using mild, olefin-modifying reagents was described and partially demonstrated for shale oil distillates and whole oils. Hydrocarbon-type composition data and a discussion of IR analysis for various types of olefinic compounds were presented for three whole oils.

Literature Cited

1. Dinneen, G. U., Smith, J. R., Van Meter, R. A., Allbright, C. S., Anthoney, W. R., *Anal. Chem.* (1955) **27**, 185.
2. Jensen, H. B., Morandi, J. R., Cook, G. L., *Prepr., Div. Pet. Chem., Am. Chem. Soc.* (1968) **13**(2), F98.
3. Morandi, J. R., Jensen, H. B., *J. Org. Chem.* (1969) **34**, 1889.
4. Jensen, H. B., Poulson, R. E., Cook, G. L., *Am. Chem. Soc., Div. Fuel Chem., Prepr.* (1971) **15**(1), 113.
5. Earnshaw, D. G., Doolittle, F. G., Decora, A. W., *Org. Mass Spectrom.* (1971) **5**, 801.

6. Morandi, J. R., Earnshaw, D. G., Jensen, H. B., Abstracts, Pittsburgh Conference on Analytical Chemistry and Applied Spectroscopy, Inc., Cleveland, Ohio, March 1972, p. 170.
7. Poulson, R. E., Jensen, H. B., Duvall, J. J., Harris, F. L., Morandi, J. R., *Proc. Ann. ISA Anal. Instrum. Symp., 18th, San Francisco, May 1972,* p. 193.
8. Allbright, C. S., Weber, J. H., Jensen, H. B., presented at the 2nd Rocky Mountain Regional Meeting of the ACS, Albuquerque, New Mexico, July 1974.
9. Jackson, L. P., Allbright, C. S., Jensen, H. B., *Anal. Chem.* (1974) **46**, 604.
10. "ASTM Book of Standards, 1975," Part 23, Method D 1019-68, p. 477.
11. "ASTM Book of Standards, 1975," Part 24, Method D 2003-64, p. 144.
12. Saier, E. L., Pozefsky, A., Coggeshall, N. D., *Anal. Chem.* (1954) **26**, 1258.
13. Duvall, J. J., Jensen, H. B., *Q. Colo. Sch. Mines* (1975) **70**(3), 187.
14. Bellamy, L. J., "The Infrared Spectra of Complex Molecules," Wiley, New York, 1962.
15. McMurry, H. L., Thornton, Vernon, *Anal. Chem.* (1952) **24**, 318.

RECEIVED October 17, 1977.

Comparative Characterization and Hydrotreating Response of Coal, Shale, and Petroleum Liquids

CARLTON H. JEWITT and GEORGE D. WILSON

Ashland Oil, Inc., P.O. Box 391, Ashland, KY 41101

A characterization and hydro-treating program utilizing liquids derived from H–Coal coal liquefaction and TOSCO II oil shale liquefaction is contrasted with appropriate counterpart petroleum streams under typical petroleum refining conditions. Studied were the naphthas (180–400°F nominal), middle distillates (400–650°F nominal), and heavy distillates (650–975°F). Naphtha hydroprocessing removed at least 95% sulfur above 600°F from shale and petroleum naphthas, no more than 90% for coal naphtha. Nitrogen removal from coal and shale naphthas was poor—50% and 55%, respectively, at 700°F. From 95% to 99% sulfur was removed from middle distillates. Denitrogenation of the middle distillates of 700°F showed 65% for coal, 51% at 700°F for shale distillate, and 96% for petroleum. Desulfurization of heavy distillates at 725°F showed 91% for coal, 83% for petroleum, and 61% for shale oil. Denitrogenation at 725°F showed 69% nitrogen removal for the petroleum fraction, 56% for the shale cut, and 17% for coal.

During 1970 and 1971, the energy industries addressed themselves to the feasibility of rapidly approaching synthetic fuels industry. Decisions were made to expend a considerable monetary and manpower effort to keep the petroleum industry abreast and informed of this new developing technology. These initial efforts resulted in decisions by Ashland Oil and a number of other corporations to participate in coal- and oil shale-conversion process development.

The coal liquefaction process from which experimental liquids were derived is the H–Coal technology (1, 2) developed by Hydrocarbon

0-8412-0395-4/78/33-170-243$05.00/1

Research, Inc., and is primarily an offshoot of their H–Oil process. In this process coal is dried, ground, and fed as a slurry to the high pressure, high temperature up-flow-ebulated bed reactor. The experimental oil shale liquids were derived from the TOSCO II process (2,3). This is an indirectly heated retorting process utilizing heated ceramic balls needed for transferring heat to the crushed oil shale to effect pyrolysis. The reactor is a rotating drum operated at just above atmospheric pressure for exclusion of air.

Generally, past practice has called for evaluating certain portions of the above liquids in existing refinery streams. This led to problems and did not allow for a detailed and comprehensive study of the very nature of the synthetic liquids. A primary purpose of this project was to attempt to define the operable, physical, and chemical nature of the liquids from coal and shale liquefaction, per se, and then contrast and compare with results from processing standard petroleum streams.

The Full-Range Liquids

The feedstocks utilized in this program were derived from a mid-continent and mid-east crude mix, from H–Coal liquefaction of No. 6 Illinois coal, and from the shale liquefaction product from the TOSCO II conversion of Parachute Creek shale. Analyses of the full-range liquids is presented in Table I. Attention is directed to the high nitrogen and

Table I. Characterization of Full-Range Crudes

	Petroleum	H–Coal	Tosco II
°API	33.5	27.6	23.1
Nitrogen, wt %	0.35	0.81	1.71
Oxygen, wt %	—	1.93	0.86
Sulfur, wt %	1.65	0.47	0.94
Carbon, wt %	—	87.6	79.8
Hydrogen, wt %	—	7.4	12.3
180°F, vol %	5.4	—	4.1
180°–400°F, vol %	15.1	17.1	15.9
400°–650°F, vol %	21.6	25.3	28.8
650°–975°F, vol %	29.3	25.3	36.9
975°F +, vol %	28.6	57.6	14.3

oxygen concentrations of the coal and shale liquids. The high resid concentration is common and expected in this particular fuel oil H–Coal mode of operation because of high coal feed space velocities. The coal liquid is high in asphaltenes and suspected deep polynuclear nitrogen–oxygen complexes. Greater than 70% of the nitrogen compounds in the coal and shale liquids are basic. The shale liquids contain high levels of metals which apparently can be concentrated into the heavy bottoms in much the same manner as is experienced in petroleum refining.

Hydro-treating

It is well known that petroleum-derived fractions, as well as coal and shale liquids, require certain upgrading through hydrogen refining depending on what end-product specifications are desired. Chemical hydrogen consumption is the best measure of determining the degree of improvement of a refinery feedstock. Hydrogen consumption during hydro-treating is primarily a function of sulfur, oxygen, and nitrogen removal and polynuclear-aromatics saturation.

This program concentrated on the requirements and problems that are expected during typical refinery processing. The main concerted efforts, therefore, were to detect and differentiate levels of performance of the synthetic liquids compared with the performances of an appropriate petroleum counterpart stream. Temperature was used as the primary process variable. If catalyst on-stream life and conversion are to be optimized, then an accurate temperature profile performance must be the initial step. Hydro-treating performances were evaluated under simulated refinery conditions while varying the most practical primary process variable, temperature. All experimental work utilized American Cyanamid HDS–3A catalyst.

The Naphthas

Analyses and characterization of the three naphtha charge stocks are presented in Table II. In addition to the boiling range, other important differences are noted in sulfur, nitrogen, and oxygen concentrations. The H–Coal naphtha contains 1800-ppm sulfur and 1307-ppm nitrogen; the TOSCO II naphtha contains 7100-ppm sulfur, 4150-ppm nitrogen, and 90-ppm phenols. These streams are compared with a petroleum naphtha containing 408-ppm sulfur, 4-ppm nitrogen, and no detectable oxygen-containing compounds.

Hydro-treating of the above naphthas took place under typical refinery conditions of 400 psig, 4 LHSV, 350 SCF H_2/bbl., and temperatures of 450°F to 700°F in 50°F increments. Desulfurization, denitrogenation, and gravity changes were measured at each temperature after the bench scale reactor unit attained equilibrium operation using a commercial nickel-moly catalyst. From 700°F the temperature then decreased in 50°F increments and the above performances were measured again to ascertain experimental integrity and any catalyst deactivation. Experimental results are detailed in Table III. Results reveal 99%

Table II.   Characterization of Naphtha Charges

	Naphtha Charges		
	Petroleum	H–Coal	Tosco Shale
°API	56.5	40.6	46.5
Sulfur (ppm)	408	1800	7100
Nitrogen (ppm)	4	1307	4150
Phenols (ppm)	—	—	90
Iron (ppm)	0.1	0.5	1.2
Nickel (ppm)	0.2	0.2	0.2
Vanadium (ppm)	0.1	0.2	0.1
Distillation (%)			
IBP	206	196	235
5	216	215	258
10	222	228	266
20	230	250	282
30	240	270	296
40	249	292	308
50	260	312	320
60	272	332	322
70	287	350	344
80	306	366	358
90	334	380	378
95	370	394	410
Saturates	88.5	70.3	N/A
Olefins	0.7	1.1	N/A
Aromatics	10.8	28.6	N/A

desulfurization of the petroleum naphtha at temperatures above 600°F. The shale naphtha showed 98% desulfurization at 700°F, but there still remained in the liquid product effluent 150-ppm sulfur, far in excess of the levels desired for reformer feed naphtha. Maximum desulfurization of the coal naphtha was accomplished at 700°F, but the liquid product still contained 170-ppm sulfur. Nitrogen removal in the petroleum naphtha was 100% at all temperatures; however, the coal naphtha underwent 48% denitrogenation at 700°F, and the shale naphtha yielded a product of 54% nitrogen removal at 700°F.

The Middle Distillates

Hydro-treating of the middle distillates must be accomplished to reduce sulfur for the obvious environmental concerns as well as to eliminate downstream processing problems. Hydrogenation of sulfur compounds over cobalt-moly or nickel-moly catalyst is generally of the first-order kinetics with respect to the sulfur compounds. Thiophenic sulfur is more difficult to hydrogenate than sulfides, disulfides, and mercaptans; the rate of hydrogenation increasing in that order. Generally, the rate of hydrogenation of the sulfur compounds decreases with increasing molecular weight of the feed, as well as that of the sulfur compounds themselves. Higher hydrogen partial pressures will, of course, enhance the hydrogenation of the sulfur compounds to effect desulfurization and

Table III. Characterization of Hydro-Treated Naphthas

	Petroleum		*H–Coal*		*Tosco Shale*	
			Naphthas			
LHSV	4		4		4	
psig	400		400		400	
Ft³/bbl H₂	350		350		350	
Catalyst bed L/D	10.9		10.9		10.9	
		% Removal		*% Removal*		*% Removal*
ppm Sulfur @ °F						
450	19	(95.4)	614	(65.9)	6700	(5.7)
500	6	(98.6)	318	(82.4)	5200	(26.8)
550	4	(99.0)	209	(88.4)	2000	(71.8)
600	9	(97.8)	182	(89.9)	300	(95.8)
650	13	(96.8)	170	(90.6)	300	(95.8)
700	15	(96.4)	170	(90.6)	150	(97.9)
650	11	(97.4)	182	(89.9)	800	(88.8)
600	4	(99.0)	180	(90.0)	2000	(71.8)
550	2	(99.5)	219	(87.8)	4300	(39.4)
ppm Nitrogen @ °F						
450	1	< (100)	834	(36.2)	3400	(25.3)
500	1	< (100)	825	(36.9)	3000	(27.7)
550	1	< (100)	773	(40.9)	2800	(32.5)
600	1	< (100)	649	(50.4)	2200	(47.0)
650	1	< (100)	692	(47.1)	2100	(49.4)
700	1	< (100)	675	(48.4)	1900	(54.2)
650	1	< (100)	705	(46.1)	2500	(39.8)
600	1	< (100)	739	(43.5)	2600	(37.3)
550	1	< (100)	743	(43.2)	3100	(25.3)
°API @ °F						
450	56.2		41.9		48.2	
500	56.2		42.4		49.9	
550	—		42.3		51.2	
600	56.2		42.6		52.2	
650	56.2		42.3		52.3	
700	55.5		42.4		52.5	
650	55.8		42.6		51.5	
600	55.9		42.1		51.3	
550	55.9		42.7		50.3	
Vol % recovery	98.0		98.5		101.2	

act also to depress the propensity for carbon formation by higher molecular weight hydrocarbons when exposed to elevated temperatures.

The middle distillates were hydro-treated under conditions of 1000 psig, 1000 SCF H_2/bbl., 3 LHSV, and temperatures ranging from 450°–700°F in 50°F increments. Again, desulfurization, denitrogenation, and gravity changes were profiled utilizing American Cyanamid HDS–3A nickel-moly catalyst.

Table IV details the characteristics of the middle distillate (400–650°F) charge stocks utilized in this study. The sulfur level of the coal liquid is low at 371-ppm sulfur, while the petroleum liquid shows 3500 ppm and the shale middle distillate analyzed at 7400-ppm sulfur. Nitrogen contents were 1800 ppm for the coal material, 11,400 ppm in the shale liquid, and 31 ppm for the petroleum. Also to be noted for the charge stocks is the wide range of gravities for similar boiling fractions. As was the case with the naphtha charges, these middle distillates contained 1.6–2.6-ppm total metals. The middle distillate charges of coal and shale contained greater than 50% of their nitrogen in the basic form.

Table V presents the tabulated experimental results of hydro-treating the middle distillates. Desulfurization results reveal at least 96% sulfur removal at 700°F for all three middle distillate materials. Nitrogen

Table IV. Characterization of Middle Distillate Charges

	Middle Distillate		
	Petroleum	H–Coal	Tosco Shale
°API	37.6	16.7	28.9
Sulfur (ppm)	3500	371	7400
Nitrogen (ppm)	31	1800	1.14 (%)
Phenols (ppm)	—	—	86
Iron (ppm)	0.5	1.0	0.6
Nickel (ppm)	0.2	0.2	0.1
Vanadium (ppm)	0.2	0.2	0.5
Distillation (%)			
IBP	440	452	436
5	460	452	454
10	474	452	470
20	486	454	485
30	494	470	500
40	504	492	515
50	514	514	530
60	522	534	548
70	532	570	566
80	546	592	582
90	560	616	606
95	574	630	616
EP	585	636	630

Table V. Characterization of Hydro-Treated Middle Distillates

	Petroleum		H–Coal		Tosco Shale	
LHSV	3		3		3	
psig	1000		1000		1000	
Ft³/bbl H₂	1000		1000		1000	
Catalyst bed L/D	10.9		10.9		10.9	
		% Removal		% Removal		% Removal
ppm Sulfur @ °F						
450	1544	(55.9)	170	(54.2)	3000	(59.5)
500	1269	(73.7)	87	(76.6)	1900	(74.4)
550	375	(89.3)	47	(87.4)	760	(89.7)
600	103	(97.1)	26	(93.0)	—	—
650	31	(99.1)	14	(96.3)	500	(93.3)
700	10	(99.7)	8	(97.9)	300	(96.0)
650	19	(99.5)	13	(96.5)	300	(79.8)
600	45	(98.8)	30	(91.9)	1500	(36.8)
550	378	(89.2)	71	(88.9)	3200	(36.8)
ppm Nitrogen @ °F						
450	—	—	—	—	—	—
500	17	(45.2)	1525	(15.3)	11300	(0.9)
550	5	(83.9)	1385	(23.1)	10200	(10.6)
600	4	(87.1)	1170	(35.0)	10030	(12.1)
650	2	(94.5)	711	(60.5)	8700	(23.7)
700	1	(96.0)	622	(65.5)	5600	(50.9)
650	1	(96.0)	881	(51.1)	5500	(51.8)
600	1	(96.0)	1274	(29.3)	7100	(37.7)
550	3	(90.4)	1481	(17.8)	7900	(30.7)
°API @ °F						
450	38.6		19.2		31.5	
500	38.4		17.3		31.6	
550	38.7		17.8		32.2	
600	39.3		18.2		34.8	
650	40.0		18.7		36.6	
700	40.4		18.9		38.7	
650	40.6		18.7		37.0	
600	39.8		17.9		36.4	
550	39.1		17.4		33.5	
Vol % recovery	98.5		102.3		101.3	

removal problems were more pronounced. The coal liquid experienced a maximum of 66% removal at 700°F compared with 51% at 700°F for the shale liquid. The middle cuts of both the coal and shale liquids contain basic nitrogen in concentrations detrimental to catalyst activity required for the full hydro-treating spectrum. A solution to this problem of basic nitrogen removal is crucial if these cuts are to be used similarly to the petroleum middle distillate streams, i.e., the charge stock for jet fuels, diesel fuels, furnace oils, and kerosene.

The Heavy Distillates

As practiced now, the polyaromatics present in gas oils, when cracked over conventional catalytic cracking catalyst, yield the paraffins and/or olefins from the alkyl side chain along with the less desirable parent polyaromatics. The net result is a product low in aromatics and high in paraffins plus olefins. Improving the properties of gas oils for cat-cracker charge can be brought about by several routes. First, the polyaromatics can be converted to monoaromatics. Increasing the monoaromatics content of the gas oil can be accomplished also by selective dehydrogenation of the naphthenes present in the original feedstock. One must appreciate, however, that this stream is a dynamic kinetic equilibrium whereby hydrogenation–dehydrogenation and hydrodesulfurization are taking place, all related to upgrading the gas oil aromatic content. As such, the total aromatics content (mono, di, tri, tetra aromatics) in the product must be related to the production (or loss) of monoaromatics content as determined in the original gas oil feedstock.

Nitrogen removal also can have a profound effect upon upgrading gas oils to improve catalytic cracking yields. The unfavorable effects of basic nitrogen compounds are thought to arise from the basic nitrogen neutralizing the active acid catalytic sites. Basic nitrogen compounds are among the worst offenders in promoting high carbon deposits on cracking catalysts. Removal of these nitrogen compounds will improve substantially cracking yields with subsequently lower off-gas and carbon make. It generally is conceded that basic nitrogen also has a detrimental effect on product distribution, octane rating, and volatility of the catalytic gasoline.

Charge stocks containing nickel and vanadium present serious processing factors. When deposited on the catalyst, they lead to light gas make, excessive coking, and pore plugging of the catalyst. These metallic contaminants cause abnormal aging with a corresponding increase in light gas and coke formation at the expense of desired liquid products.

The heavy distillate fractions of shale and coal liquids will require hydro-treating in the event that they were to be used as an FCC charge stock or a fuel oil. The hydro-treating of these heavy distillates does not appear to facilitate optimum application of these materials; however, this program's primary objective was to study the hydro-treating of the complete full range distillates and their subsequent comparison with petroleum.

These heavy gas oils were processed under conditions of 1300 psig, 3000 SCF H_2/bbl., 1 LHSV, and temperatures of 625°, 675°, and 725°F. In addition to sulfur and nitrogen removals, hydrogen consumption and aromatics saturation monitoring was attempted.

The characterization of the heavy distillate charges is contained in Table VI. Sulfur levels were 9700 ppm for petroleum, 1500 ppm for the coal liquid, and 5900 ppm for the shale. Nitrogen again is a paramount problem with a 3600-ppm concentration in the coal heavy distillate and 1.04 wt % in the shale liquid. This contrasts to 303 ppm nitrogen in the petroleum material. Total metals were 3.3 ppm for the petroleum heavy cut, 17.2 ppm for the coal material, and 478-ppm total metals in the shale distillate.

Table VI.   Characterization of Heavy Distillate Charges

	Heavy Distillate		
	Petroleum	H–Coal	Tosco Shale
°API	27.9	5.4	12.6
Sulfur (ppm)	9700	1500	5900
Nitrogen (ppm)	303	3600	1.04 (%)
Phenols (ppm)	—	—	—
Iron (ppm)	1.3	15.3	80
Nickel (ppm)	0.2	0.2	4.8
Vanadium (ppm)	0.6	0.2	1.7
Distillation (%)			
5	493	682	738
10	531	688	761
20	597	699	803
30	641	706	850
40	692	722	894
50	737	737	942
60	789	756	982
70	841	783	989 (%)
80	901	843	cracked
90	973	896	cracked
95	1011	944	cracked

Table VII presents the results of hydro-treating the heavy distillates. The petroleum fraction was 56% desulfurized at 625°F and 83% at 725°F. Desulfurization of the coal material was 82% at 625°F and 90% at 725°F. The shale liquid underwent 28% desulfurization at 625°F and 56% at 725°F. Denitrogenation of the coal and shale heavy gas oils was very poor. Maximum denitrogenation occurred at 725°F and gave values of 69% for petroleum, 17% for coal, and 56% for the shale material. The tabulated experimental results are presented in Table VII.

In conclusion, based upon results of this program, several general observations are apparent. The naphthas of both coal and shale lique-faction require further process development studies if these liquids are to be included in charge stocks for multimetallic catalytic reforming. This includes studies of space velocity reductions, higher hydrogen partial pressures, and increased hydrogen circulation rates, either singly or in combination. The light gas oils appear to present minimum prob-lems for conventional refinery processing techniques. The heavy gas oils appear best suited for end uses other than catalytic cracking charge stock, at least for the present. The problems encountered and expected though many, surely are not insurmountable based on the achievement records and past performance of the petroleum industry.

Table VII. Characterization of Hydro-Treated Heavy Distillates

Heavy Distillate

	Petroleum		*H–Coal*		*Tosco*	
LHSV	1		1		1	
psig	1300		1300		1300	
Ft^3/bbl H$_2$	3000		3000		3000	
Catalyst bed L/D	6		6		6	
		% Re-moval		*% Re-moval*		*% Re-moval*
ppm Sulfur @ °F						
625	4300	(55.7)	277	(81.5)	4500	(23.8)
675	2100	(78.4)	239	(84.1)	3600	(39.0)
725	1700	(82.5)	134	(91.1)	2300	(61.1)
ppm Nitrogen @ °F						
625	182	(40.0)	3700	—	7500	(27.9)
675	122	(59.7)	3500	(2.8)	5900	(43.3)
725	94	(69.0)	3000	(16.7)	4600	(55.8)
°API @ °F						
625	31.6		9.2		17.0	
675	33.1		9.5		17.5	
725	33.7		9.9		19.9	

Summary

Characterization of the typical full range liquids reveals oxygen concentrations of 1.9 wt % in the H–Coal liquid and 0.9 wt % in the TOSCO shale liquid. Processing problems are expected with nitrogen removal because greater than 70% of the nitrogen compounds are basic in both the coal (0.8 wt % total nitrogen) and shale liquids (1.7 wt % total nitrogen). Sulfur concentrations of 0.5 wt % in coal liquids and 0.9 wt % in the shale liquids are moderate, but present problems caused by the large percentage of sulfur associated with multi-ring heavy aromatic compounds in both the coal and shale liquids. Metal contaminants appear to present problems with the 400+ fractions of both coal and shale liquids.

Generally, the program consisted of the hydro-treating of naphthas, middle distillates (400°–650°F), and heavy distillates (650°–975°F) derived from coal liquefaction (H–Coal), shale oil liquefaction (TOSCO II), and petroleum.

The naphthas were hydro-treated under conditions of 400 psig, 4 LHSV (liquid hourly space velocity), 350 SCF H_2/bbl (standard cubic feet hydrogen per barrel), and temperatures of 450°–700°F in 50°F increments. Desulfurization, denitrogenation, and gravity changes were measured at each temperature after the unit attained steady state equilibrium operation. The catalyst utilized is a commercial nickel-moly on alumina. Results of the naphtha hydroprocessing reveal 95% and greater sulfur removal above 600°F for the shale and petroleum naphthas. The coal naphtha did not experience greater than 90% desulfurization. Nitrogen removal in both the coal and shale naphthas was very poor, reaching a maximum of 50% and 55%, respectively, at 700°F. These nitrogen levels cannot be tolerated by any of the current multimetallic reformer catalysts if these naphthas are to be included in reformer charge stocks.

The middle distillates were processed under conditions of 1000 psig, 1000 SCF H_2/bbl, 3 LHSV, and temperatures ranging from 450°–700°F. Again, desulfurization, denitrogenation, and gravity changes were profiled carefully. Desulfurization of the middle distillates revealed parallel performances for the three liquids, all attaining 95%–99% sulfur removal. Denitrogenation of these streams again proved difficult; 65% at 700°F for the coal middle distillates and 51% at 700°F for the shale middle distillates.

The heavy gas oils were processed at 1300 psig, 3000 SCF H_2/bbl, 1 LHSV, and at temperatures of 625°, 675°, and 725°F. Desulfurization efficiency of the coal heavy distillate was greater than that for the petroleum. Results show that at 725°F the coal heavy distillate was 91% desulfurized, the petroleum fraction was 83% at 725°F, and the shale heavy distillate analyzed at 61% sulfur removal. Here again, the nitrogen compounds proved very difficult to remove. At 725°F the petroleum fraction underwent 69% nitrogen removal, the shale cut revealed a 56% denitrogenation, and the coal heavy distillate exhibited 17% introgen removal.

Acknowledgment

The authors wish to express their gratitude to Ashland Oil, Inc. and Hydrocarbon Research, Inc. for their cooperation in this project.

Literature Cited

1. Johnson, C. A., et al., "Scale-Up Factors in the H-Coal Process," *Chem. Eng. Prog.* (1973) **69** (3), 52–54.
2. Hendrickson, T. A., "Synthetic Fuels Data Handbook," Cameron Engineers, Inc., Denver, 1975.
3. Lenhart, A. F., "The TOSCO Process–Economic Sensitivity to the Variables of Production," *Midyear Meeting, Am. Pet. Inst., Am. Inst. Min., Metall. Pet. Eng., 34th, 1969.*

RECEIVED August 5, 1977.

High-Precision Trace Element and Organic Constituent Analysis of Oil Shale and Solvent-Refined Coal Materials

J. S. FRUCHTER, J. C. LAUL, M. R. PETERSEN,
P. W. RYAN, and M. E. TURNER

Physical Sciences Department, Battelle, Pacific Northwest Laboratories,
Richland, WA 99352

Materials from a solvent-refined coal pilot plant and two simulated in situ oil shale retort facilities have been characterized for trace inorganic and organic compounds. The techniques used allowed the determination of some 30 elements, the chemical and physical forms of arsenic and mercury, and a large number of organic compounds. Satisfactory balances were obtained for most trace elements except mercury in effluents from the solvent-refined coal plant and one of the oil shale retorts. Approximately 60 organic compounds were determined quantitatively in process streams from the solvent-refined coal plant, and 20 organic compounds were determined in the crude shale oil from an oil shale retort pilot plant.

Oil shale and coal conversion technologies are presently in a stage of rapid development. Analysis of materials involved in these processes is an important part of the investigation of possible environmental and health impacts of the processes. High precision analyses are desirable even in those cases where sampling uncertainties are relatively large, so that the analytical procedures will not add appreciably to the overall error.

In this paper, the application of a number of sensitive and precise methods for the determination of trace elements, heavy element species, and organic compounds in materials from one oil shale research retort process and from a solvent-refined coal pilot plant operation are discussed. The methods used were chosen both for their sensitivity and their relative freedom from interferences.

0-8412-0395-4/78/33-170-255$06.75/1
© 1978 American Chemical Society

Samples. The Laramie Energy Research Center (LERC) and the Lawrence Livermore Laboratory provided the oil shale materials from their simulated in situ research retorts. The oil shale obtained from LERC was from the Piceance Creek Basin of Colorado and had reported a Fischer assay of ~ 15 gal of oil per ton. Spent oil shale, oil, and water samples also were obtained but are not from the same retort runs. Samples of offgases were taken from LERC's 10-ton retort. Samples of 24 gal/ton raw shale (Anvil Points mine), spent shale, retort water, and crude oil were obtained from Livermore's 125-kg retort (Run S-11). A miscellaneous shale sample from the Piceance Basin with a Fischer Assay of ~ 40 gal/ton was analyzed also. In view of the variability of oil shale from different sites and the differing characteristics of different retort processes, the results of the analyses should not be generalized.

Coal materials were provided by Pittsburg and Midway Coal Mining Company, which operates the solvent-refined coal (SRC) pilot plant at Ft. Lewis, Washington. They made available to us feed coal (Kentucky high-sulfur bituminous), solvent-refined coal, mineral residue, light oil, wash solvent, process solvent, effluent water, holding pond sediment, and stack particulates. Since the SRC plant is a pilot operation, the experimental operating conditions and materials are changed frequently. Therefore, we have found that analytical results based on individual samples may vary substantially between plant runs.

Analytical Techniques—Inorganic

Neutron Activation. In environmental and industrial process studies, neutron activation analysis (NAA) currently is being used widely because of its inherently high sensitivity and accuracy. In complex substances such as are found in solvent-refined coal and oil shale retorting processes, NAA is the method of choice for many trace element analyses because of its relative freedom from matrix effects.

The NAA method can be divided into NAA (Instrumental NAA) and RNAA (Radiochemical NAA). In the latter, the various neutron-induced products are separated chemically to minimize interferences. There are several comprehensive review papers on INAA published in the literature (1, 2, 3, 4). Briefly, the basic parameters controlling sensitivity for a multi-element determination are neutron flux, irradiation time, delay interval prior to counting, half-life and gamma-ray energy of the induced activity, and efficiency and resolution of the detector. Table I outlines the irradiation parameters used for each of the two sequential irradiations. The final count occurring 40–50 days after the second irradiation is performed on an anti-coincidence-shielded Ge(Li) system developed recently in our laboratory.

Table I. Instrumental Neutron Activation Analysis (n,γ) Process[a]

Isotope	Half-Life	γ-Energy Selected (keV)	Decay Interval	Possible Interference
I. ^{51}Ti	5.79 min	320	15 min	
^{27}Mg	9.46 min	1014	15 min	^{27}Al(n,p)
^{52}V	3.75 min	1434	15 min	
^{28}Al	2.32 min	1779	15 min	^{28}Si(n,p)
^{49}Ca	8.80 min	3084	15 min	
^{165}Dy	2.32 hr	95	3–5 hr	
^{56}Mn	2.58 hr	847	3–5 hr	^{56}Fe(n,p)
^{24}Na	15.0 hr	1369	3–5 hr	^{24}Mg(n,p), ^{27}Al(n,α)
II. ^{153}Sm	46.8 hr	103	5–7 days	
^{152}Eu	12.7 yr	122	5–7 days	
^{177}Lu	6.74 days	208	5–7 days	^{177m}Lu(208)
^{239}Np	56.3 hr	228	5–7 days	^{177m}Lu(229)
^{175}Yb	4.21 days	396	5–7 days	
^{131}Ba	12.1 days	496	5–7 days	
^{147}Nd	11.1 days	531	5–7 days	
^{140}La	40.2 hr	816,1597	5–7 days	
^{24}Na	15.0 hr	1369,1733	5–7 days	^{24}Mg(n,p), ^{27}Al(n,α)
^{76}As	26.3 hr	559	5–7 days	
^{42}K	12.4 hr	1524	5–7 days	
^{182}Ta	115 days	68,1221	40–50 days	
^{152}Eu	12.7 yr	122,1408	40–50 days	
^{141}Ce	32.5 days	146	40–50 days	^{59}Fe(143)
^{160}Tb	72.1 days	299	40–50 days	^{233}Pa(300)
^{233}Pa(Th)	27.0 days	312	40–50 days	
^{51}Cr	27.8 days	320	40–50 days	
^{181}Hf	42.5 days	482	40–50 days	
^{85}Sr	64.0 days	514	40–50 days	
^{95}Zr	65.5 days	757	40–50 days	^{154}Eu(757)
^{134}Cs	2.05 yr	796	40–50 days	
^{58}Co(Ni)	71.3 days	811	40–50 days	^{152}Eu(811)
^{46}Sc	83.9 days	889,1121	40–50 days	
^{86}Rb	18.7 days	1078	40–50 days	
^{59}Fe	45.6 days	1099,1292	40–50 days	
^{65}Zn	243.0 days	1116	40–50 days	^{152}Eu(1112), ^{46}Sc(1121)
^{60}Co	5.26 yr	1173,1332	40–50 days	
^{214}Sb	60.3 days	1691	40–50 days	
^{75}Se	120 days	265,280,400	40–50 days	^{203}Hg(280)

[a] Parameters: neutron flux, 6×10^{12} n/cm^2/sec; irradiation I, 5 min; irradiation II, 6 hr.

In brief, when an isotope emits two or more γ-rays in cascade, and if they are detected simultaneously by the NaI(Tl) anti-coincidence shield and the Ge(Li) diode, the coincidence events are stored in one-half of a 4096-channel memory, while the single γ-ray or events in non-coincidence are stored in the second half of the memory. Figure 1 displays coincidence and non-coincidence spectra of the geological standard rock BCR-1. The great advantage in non-coincidence lies in the fact that the Compton continuum (a background interference) is reduced by an order of magnitude for low and medium γ-ray energies. Thus, peak/Compton ratios and sensitivities for many elements such as chromium (Cr), strontium (Sr), barium (Ba), neodymium (Nd), rubidium (Rb), zinc (Zn), selenium (Se), and antimony (Sb) in non-coincidence spectra, usually low in normal Ge(Li) counting, are increased greatly. The 1116-keV peak of Zn was measured easily by the reduced-Compton edge of the 1121-keV peak of scandium (^{46}Sc) in non-coincidence. The 146-keV peak of cerium (^{141}Ce), unlike in normal Ge(Li) counting, contained relatively little contribution from the 143-keV peak of iron (^{59}Fe). The elements nickel (Ni) via cobalt (^{58}Co) and zirconium (Zr) not observed previously in our normal Ge(Li) counting were easily measurable in non-coincidence counting. The 757-keV peak of Zr and the 811-keV peak of Ni have interferences from europium (^{152}Eu). The contribution amounted to 0.71% for the Zr peak and 1.0% for the Ni peak relative to the 122-keV peak of ^{152}Eu. Overall, the accuracy and relative precision for a number of elements was improved by factors of 3–10 by non-coincidence counting relative to the normal Ge(Li) counting.

Radiochemical group separation has a major advantage over individual element separation in that it is far less time consuming, yet it can permit sufficient separation to allow precise analysis. For example, by simply separating the rare earth elements (REE) as a group after neutron activation, it is possible to measure most of the rare earth spectra by direct counting and thus determine their distribution.

Our procedure involves irradiation of samples and standards to a total neutron exposure of 1×10^{17} n/cm². After a two-day decay, the samples were transferred into Ni crucibles already containing the dried REE carriers. The samples were fused using a Na_2O_2–NaOH mixture. The fusion cake was decomposed with distilled water and neutralized with HCl. The REE were precipitated as a group with NH_4OH, dissolved in HCl, and precipitated as fluoride with NH_4HF_2 and HF. The REE fluoride was dissolved in H_3BO_3 and HNO_3 and precipitated as REE hydroxide with NH_4OH. The fluoride–hydroxide cycle was repeated three times to insure radiochemical purity.

The rare earth group aliquots were counted on a Ge(Li) system. The first count was made after three days. After a decay of 20–30 days,

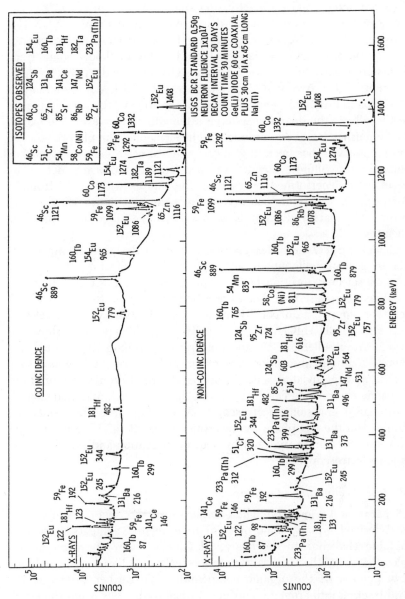

Figure 1. Anti-coincidence-shielded Ge(Li) γ-ray spectra of irradiated U.S.G.S. Standard Basalt BCR-1

the samples were counted again for 100–1,000 min to optimize the detection sensitivities for long-lived isotopes of Ce (cerium), Nd, Eu (europium), Gd (gadolinium), Tb (terbium), and Tm (thulium). Chemical yields were determined by reactivation, and ranged from 50–80%.

X-Ray Fluorescence. Energy-dispersive x-ray fluorescence (XRF) spectroscopy is used to measure certain trace elements where greater sensitivity or accuracy can be obtained by this method. Some trace elements are measured by both NAA and XRF to provide a check on the consistency of the two analytical methods. The XRF determination of low Z (atomic number) elements is quite sensitive to the sample matrix since this affects the degree of photon attenuation. However, the matrix effect diminishes for higher atomic weight elements. Sample preparation technique for XRF has been described previously (5). Sample standards are mixed with cellulose powder and pressed into uniform pellets. The XRF analysis is performed by exciting the sample with x-rays from Zr or Ag (silver) targets and observing the x-ray emission with a Si(Li) detector. The data reduction used has been described in a previous publication (6).

Chemical Speciation Measurements. Our analysis of the trace elements in gaseous effluents is accomplished by collecting the various species on selected sorption beds within a gas sampling train. Beds for Hg and As (arsenic) have been described in the literature (7, 8, 9). A typical Hg sample train consists of a series of traps which selectively absorb $HgCl_2$, CH_3HgCl, $Hg°$, and $(CH_3)_2Hg$. This train is shown in Figure 2. After collection, the sorption traps containing Hg are analyzed by flameless atomic absorption spectroscopy. Similarly, arsenic as arsine, the oxide and methylated species can be absorbed on a series of traps. The arsenic compounds can be removed then, reduced, and analyzed by DC discharge emission spectroscopy. A number of trace metals are determined in the gas phase by absorption in acidic and basic impinger solutions which can be analyzed by neutron activation and DC discharge emission spectroscopy.

Analytical Techniques—Organic

Separation Techniques. The complexity of the organic composition of coal-derived liquids, shale oil, and their related effluents presents a formidable challenge to the analytical chemist. Our approach to this problem has been the classical separation technique based on acid–base–neutral polarity of the organic compounds. We further subdivide the neutral fraction into aromatic and non-aromatic fractions using dimethylsulfoxide (DMSO) extraction. DMSO effectively removes multiringed aromatic compounds with great efficiency (85–95%) for these complex mixtures and thus allows a straightforward analysis for polynuclear

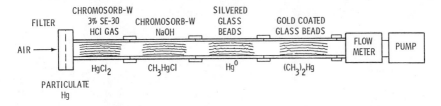

Figure 2.   Selective adsorption traps for mercury species in gases

aromatic compounds. The overall extraction scheme is depicted in Figure 3 for shale oil. Any liquid mixture containing organic constituents may be treated similarly and the components determined by any suitable chromatographic means.

There are disadvantages in this extraction scheme. Volatile and water-soluble compounds are lost during evaporation of the organic phase or insufficiently back-extracted and discarded with the aqueous layer. With some mixtures such as shale oil a third phase is present which is a tar layer soluble in neither the organic nor the aqueous layer. Sludge formation occurs during work-up, indicating that reactions are taking place during the extraction. For these reasons we feel that this extraction scheme is not the final answer. However, we initially chose to use this scheme because of its operational simplicity and utility. As more information becomes available about the fractions, both in chemical content and biological activity (e.g. Ames test results), we intend to develop more sophisticated schemes.

Methods for Particulates and Gases. For determining the organic content of particulates we have found the usual method of collecting airborne particulates on specially cleaned fiberglass filters to be suitable. The particulate filters then are extracted and the organic extract can be further partitioned by the above scheme or analyzed without any separation.

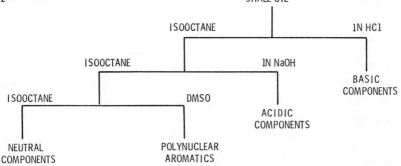

Figure 3.   Separation scheme for shale oil

Table II. Analyses of Samples

Samples	Light Oil	Wash Solvent	Process Solvent
		PNA Fraction	
Xylene		1300	
o-Ethylbenzene	9800	1700	
m/p-Ethylbenzene		700	
C$_3$-Benzene	3900	1500	
C$_4$-Benzene		500	
Indan	4300	13000	
Methylindan	510	2500	
Methylindan	180	1400	
Methylindan	240	2300	
Dimethylindan	< 5	40	
Tetralin	330	4100	
Dimethyltetralin	< 5	1500	
6-Methyltetralin	110	3200	
Naphthalene	1630	32000	100
2-Methylnaphthalene	690	32000	3800
1-Methylnaphthalene	110	12000	930
Dimethylnaphthalene	80	13000	11200
Dimethylnaphthalene	70	700	1700
Dimethylnaphthalene		4000	4200
Dimethylnaphthalene	10	160	650
2-Isopropylnaphthalene		40	50
1-Isopropylnaphthalene		210	1400
C$_4$-Naphthalene		5	50
Cyclohexylbenzene		410	
Biphenyl	80	10000	5900
Acenaphthylene	2	500	3400
Dimethylbiphenyl			
Dimethylbiphenyl	15	35	2100
Dimethylbiphenyl	21	30	560
Dibenzofuran	8	400	5800
Xanthene	10	30	840
Dibenzothiophene	3	50	4200
Methyldibenzothiophene		15	320
Dimethyldibenzothiophene	5	15	1200
Thioxanthene		2	3300
Fluorene	15	250	6600
9-Methylfluorene	15	110	3100
1-Methylfluorene	10	10	3000
Anthracene/phenanthrene	25	130	23000
Methylphenanthrene	6	15	6200
1-Methylphenanthrene	6		3900
C$_2$-Anthracene	6	25	500
Fluoranthene	15	35	10500
Dihydropyrene	6	25	1200
Pyrene	20	40	11200

Collected at Solvent-Refined Coal Plant[a]

Raw Process Water	Mineral Residue	Solvent-Refined Coal	Particulate Filter (Concentration $\mu g/m^3$)
		PNA Fraction	
	85		
15	25		
	55		
	25		
< 0.1	110		
0.5	35		
	50		
5	1500	1	3
2	740	8	16
	180	5	4
0.3	260	6	170
	60	3	20
	150		
2	< 2		10
0.7	< 1		< 0.5
2	15		20
	< 1		4
	5		
0.2	270	2	75
< 0.1	45	8	60
0.5	30	9	130
0.2	20	7	40
0.6	60	9	160
0.1	20	5	40
1.5	70	30	180
< 0.1	8	4	60
< 0.05	20	13	130
0.1	5	3	120
0.3	80	27	200
0.3	40	11	150
0.2	50	18	100
1.1	500	300	1500
0.3	100	50	400
0.2	50	30	300
< 0.05	10	1	30
0.4	200	180	700
< 0.05	10	1	30
0.6	200	280	900

Table II.

Samples	Light Oil	Wash Solvent	Process Solvent
Neutral Fractions (n-alkanes)			
n-Octane	16000	900	
n-Nonane	8700	2700	
n-Decane	9800	5000	
n-Undecane	3900	8300	50
n-Dodecane	1400	21000	80
n-Tridecane	470	14000	30
n-Tetradecane	170	11000	340
n-Pentadecane	60	4000	1000
n-Hexadecane	10	400	2000
n-Heptadecane	10	120	3100
n-Octadecane		40	920
n-Nonadecane		500	800
n-Eicosane			930
n-Heneicosane			600
n-Docosane			670
n-Tricosane			980
n-Tetracosane			900
n-Pentacosane			740
n-Hexacosane			450
n-Heptacosane			300
n-Octacosane			150
n-Nonacosane			90
n-Triacontane			60
n-Hentriacontane			40
n-Dotriacontane			10
n-Tritriacontane			5

The Following Compounds Have Been Identified But

PNA's

Methylpyrenes
Benzofluorenes
C_2-Pyrene
C_2-Fluoranthene
Tetrahydrochrysene
Chrysene
Methylbenzofluorene
C_3-Pyrene
C_3-Fluoranthene
Methylchrysenes
Methylbenzanthracenes
Cholanthrene
Tetrahydrobenzofluoranthene

Continued

Raw Process Water	Mineral Residue	Solvent-Refined Coal	Particulate Filter (Concentration $\mu g/m^3$)
		Neutral Fraction (n-alkanes)	
2.3			
0.3	90	4	
0.3	550	10	
0.4	9100	8	4
0.3	210	7	12
0.2	80	12	18
0.2	50	8	50
0.02	20	3	35
	10	3	18
	16	22	30
	14		20
	14		35
	16		55
	14		35
	14		45
	10		43
	8	5	40
	6	2	25
	5	2	28
	4	1	18
	2	1	22
	1	1	15
	< 1		11
			7

Have Not Yet Been Quantitatively Determined:

Phenols	Bases
Phenol	Tetrahydroquinolines
o-Cresol	Methyltetraheydroquinoline
m/p-Cresol	Quinoline
2,6-Xylenol	Isoquinoline
2,5-Xylenol	Methylquinolines
2,4-Xylenol	Dimethylquinolines
3,5-Xylenol	Ethylquinolines
o-Ethylphenol	Indole
m/p-Ethylphenol	Methylindoles
Trimethylphenols	Dimethylindoles
Methylethylphenol	Acridine
	Methylacridines
	Carbazole

Table II.

PNA's

Tetrahydrobenzopyrenes
Benzopyrenes
Methylbenzopyrenes
Methylbenzofluoranthenes
Benzofluoranthenes

[a] Concentration (ppm).

The organic constituents of gaseous materials are determined either directly by injection of suitable aliquots of the mixture (up to 5 mL) into a gas chromatograph equipped with subambient temperature programming, or by adsorption onto polymer traps, e.g. Tenax. The direct method is usually suitable for organic compounds in concentrations of more than 1 ppm such as occur in process gas streams, while the adsorption method is more appropriate in trace gas analyses.

High-Pressure Liquid Chromatography. In our experience, high-pressure liquid chromatography (HPLC) is the method of choice for looking at effluent waters in the fuel conversion processes. The method is especially suited for volatile organic compounds and for water-soluble compounds. For example, phenols in SRC waters or light N-heterocyclics in shale retort waters are determined readily at ppb levels using reverse phase techniques with UV detection. Because of the complexity of the crude coal and shale oil liquids, we feel HPLC is not suitable for component analysis because of the lack of adequate detectors or chromatographic resolution of present HPLC technology.

Gas Chromatography and Gas Chromatography–Mass Spectrometry. The extracts and the separated extracts—acidic–basic–neutral and poly-nuclear aromatic—were analyzed using gas chromatography (GC) and gas chromatography–mass spectrometry (GC–MS). A variety of GC–MS techniques were used to identify all major components and many minor ones, leading to a fraction-by-fraction, component-by-component characterization of each sample from the coal-derived liquids as given in Table II. Efficient packed columns have served well in this characterization; however, as more information is needed on the minor components, we are turning more to glass capillary columns to obtain the necessary resolution.

Of course, one goal in analytical work is to increase the accuracy and precision of the measurements. Hence a major objective of our GC–MS effort has been to develop simple, rapid techniques which can be applied to these extremely complex materials with a minimum of preliminary separation and concentrations. Various techniques of chemical ionization

Continued

Phenols	Bases
	Methylcarbazoles
	Benzoquinolines
	Methylbenzoquinolines
	Benzoindoles
	Methylbenzoindole

mass spectrometry (CI) are suited to this effort because CI often produces simple characteristic spectra which are easily distinguishable from interferences. It is often possible to find CI conditions which selectively ionize a given class of compounds while suppressing mass spectral contributions from others. For example, normal paraffins are the most abundant class of compounds in shale oil, but these compounds produce no ions when isobutane is used as a CI reagent gas. This phenomenon allows observation of mass spectra of nitrogen heterocyclic species and other less abundant species free from interference by the paraffins. We also have used a charge exchange technique with argon as the CI reagent. This technique produces large molecular ion currents from polynuclear aromatic hydrocarbons (PNA's) while suppressing interferences from other compounds. This technique is possible because most ions of a mass comparable to PNA fragment into low mass ions which can be ignored. Sub-ppm concentrations of benzopyrenes and other PNA's often can be measured in a sample directly with no required sample preparation. By using a 1-ft 3% SP2100/0.3% SP1000 column we can separate the benzopyrenes from other PNA's with an analysis time of less than 5 min. Concentrations of benzopyrenes in SRC by-products, two shale oils, and a crude petroleum which were determined by this method are given in Table III.

Table III. **Benzopyrenes Determined in Solvent-Refined Coal and Oil Shale Materials**

	ppm
SRC By-Products	
Naphtha	0.2
Wash Solvent	0.2
Recycle Solvent	360
Laramie	
150 T (Run 13)	23
Livermore	
(Run S-11)	3
Prudhoe Bay Crude	0.3

Inorganic Results

Oil Shale Retort Offgas. Table IV gives the chemical and physical forms of As and Hg in a gaseous effluent from retorting a batch of oil shale in the 10-ton retort at Laramie. The gas samples were obtained both before and after the offgas burner, which serves as an emission control device. Most of the As before the offgas burner in this retort run was in the form of arsine gas. After the offgas burner, a substantially larger fraction of the As was trapped on a nucleopore filter and appears to be in the form of small As trioxide particles. Hg was largely in gaseous form, either as elemental Hg or dimethyl Hg with smaller amounts of other forms.

Oil Shale and Solvent-Refined Coal Liquids. Measurable amounts of the heavy elements Co, Hg, As, Zn, and Se were present in both crude shale oil and in the coproduced water from the Livermore Retort Run S-11 and the Laramie 150 T Retort Run 13. In addition, the process waters from both retorts contained significant Br (bromine) and Sb, and the water from the Laramie retort contained significant uranium (U). The data for the oil and water are presented in Tables V and VI. Independent data for several elements in the Livermore S-11 oil, water shale, and split shale can be found in Reference 10. The results for process solvent and light oil from the solvent-refined coal plant are shown in Table VII. These two liquids were generally low in trace elements, although the process solvent contained some Zn and the light oil contained measurable amounts of Zn, Br, Cr, and As. Estimated precisions for the elements shown in Tables V, VI, and VII are given in Table VIII. Independent data for a number of elements in solvent-refined coal materials can be found in Reference 11.

Table IV. Trace Elements and Species in Offgas from the LERC 10-Ton Research Oil Shale Retort

	Before Offgas Burner (ng/m^3)	After Offgas Burner (ng/m^3)
Arsenic		
Gaseous	15,000[a]	1400
Particulate	400	2500
Mercury		
Particulate	150	600
Hg^{++}	1,000	100
$(CH_3)Hg^+$	800	300
$Hg^0 + (CH_3)_2Hg$	3,000	900

[a] Occurs mostly in the form of arsine (AsH_3).

Table V. Trace and Minor Elements in Oil Shale Retorting Materials from the Lawrence Livermore Laboratory's 125-kg Simulated in Situ Retort (in ppm) Average of 4 Splits

Element	Raw Oil Shale (Anvil Points 24 gal/ton) NAA	Spent Oil Shale[a] (S-11) NAA	Crude Shale Oil (S-11) NAA	Process Water (S-11) NAA
Na	1,620 ± 300	2410	26 ± 1	52 ± 1
Fe	23,300 ± 1000	32,400	57 ± 4	14 ± 1
Cr	34 ± 4	50	0.27 ± .10	0.08 ± .007
Co	8.6 ± 0.1	11.8	0.35 ± .08	0.023 ± .002
Sc	6.4 ± .02	9.8	0.0004 ± .00006	0.0003 ± .00005
Ba	560 ± 60	770	< 0.5	< 0.3
La	20.3 ± 0.4	30.0	< 0.003	< 0.003
Sm	3.5 ± .04	5.2	< 0.003	< 0.002
Eu	0.59 ± .02	0.86	< 0.001	< 0.001
Hf	1.9 ± .3	2.8	< 0.002	< 0.002
Th	6.0 ± 0.2	9.9	< 0.0006	0.004 ± .0008
Rb	75 ± 3	107	< 0.07	0.07 ± .01
As	54 ± 2	59	18 ± 2	4.1 ± .06
Sb	2.1 ± 0.1	2.9	.038 ± .06	.047 ± .03
Zn	85 ± 3	110	5.6 ± 0.6	10.0 ± .05
Br	< 0.9	< 0.9	0.2 ± 0.1	0.1
Se	1.9 ± .2	1.5	2.7 ± .4	0.77 ± .05
Hg[b]	0.08 ± .01	0.008 ± .002	0.45 ± .03	0.024 ± .005

[a] Estimated precision same as raw oil shale.
[b] Determined by Flameless Atomic Absorption Analysis.

Oil Shale and Solvent-Refined Coal Feedstocks and Solid Effluents. Data for spent oil shales and raw oil shale samples are shown in Tables V and VI. These raw shales and the spent shale from the Laramie retorts shown in Table VI were not from the same run so that they cannot be compared directly. The trace elements in the raw shales did not differ significantly. Table V shows that the spent shales are somewhat richer in most nonvolatile trace elements than in raw shales. This is expected because the volatile organic material that was removed was quite low in these elements. The more volatile trace elements were somewhat lower in spent shale. Data for solvent-refined coal materials are shown in Table VII. Table IX shows the factors by which the various elements were depleted in solvent-refined coal as compared with the feed coal. Most of the refractory trace elements, as well as As, Ni, Sb, Pb, and Se, were depleted by relatively large factors. These elements remained with the mineral residue.

Calculated Material Balances. Material balances calculated for selected elements are shown graphically in Figures 4 and 5 for the Livermore 125-kg retort and the SRC pilot plant. The mass flow balances

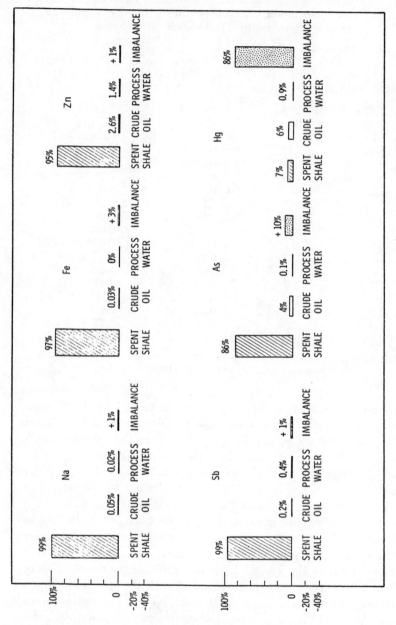

Figure 4.   Calculated material balances for selected elements in the Livermore simulated in situ oil shale retort, Livermore, California, Run S-11

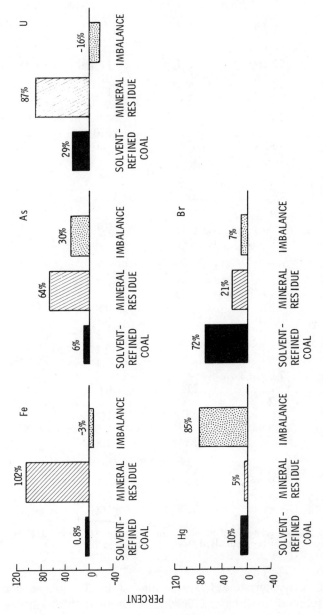

Figure 5. Calculated material balances for selected elements in the SRC pilot plant, Ft. Lewis, WA

Table VI.   Trace and Minor Elements in Oil Shale Retorting

Element	Raw Oil Shale 15 gal/ton		Raw Oil Shale 40 gal/ton	
	NAA	XRF	NAA	XRF
Na	16,100			
Fe	20,700	20,700	17,400	
Cr	30		21	
Co	11		11	
Ni		33	30	
Sc	5.3		4.9	
Ba	320		370	
Sr	440	380	490	
La	15.9		16.2	
Ce	31		30	
Pr	3.9			
Nd	15			
Sm	2.4			
Eu	0.43		0.45	
Gd	1.9			
Tb	0.26			
Ho	0.33			
Er	1.0			
Tm	0.14			
Yb	0.87		0.8	
Lu	0.140		0.18	
Hf	0.71		1.2	
Ta	0.53		0.44	
Th	4.3		4.4	
U	7.0		5.6	
Rb	66	71	63	
Cs	5.2		5.2	
As	81	63	108	
Sb	3.5		3.6	
Zn		136	100	
Br				
Se		43	6	
Hg[a]	0.2		0.1	
Ca		4.1	5	
Cu		72	56	

[a] Where estimated precision is not shown in this table, see Table VIII.
[b] U. S. Geological Survey Standard Basalt BCR-1.

used for the Livermore retort were supplied by Livermore personnel and are as follows: raw shale, 125.2 kg; spent shale, 85.6 kg; dry shale oil, 10.54 kg; and retort water, 3.77 kg. The mass flow balance for the (SRC) process was provided by Pittsburg and Midway Coal Company personnel and was as follows: feed coal, 100%; SRC, 56%–63%; mineral

Materials from the Laramie Simulated in Situ (in ppm) Retorts[a]

Spent Oil Shale		Crude Shale Oil (Run 13)	Process Water (Run 13)	BCR[b]
NAA	XRF	NAA	NAA	NAA
23,400		19.4 ± 0.4	98 ± 1	23,500
42,100	47,000	30 ± 1	< 1	95,900
52		0.04 ± 0.01	< 0.02	15
21.1		0.37 ± 0.02	0.65 ± 0.004	36
	70			16
9.4		< 0.0002	0.0004 ± 0.0001	32
650		< 0.014	0.13 ± 0.02	680
1120	1080	< 0.090	0.150 ± 0.05	320
34		0.0009	0.0017	24.6
62		< 0.009	0.100	52
7.5		< 0.0013	< 0.0006	7.2
30		< 10	0.055	30.3
4.90		0.00009	0.00031	6.3
0.82		< 0.0002	0.00029	1.95
3.3		< 0.00065	0.02	6.90
0.51		0.00006	0.0011	1.01
0.78		< 0.0002	0.00005	1.26
2.0		< 0.008	< 0.0015	3.4
0.36		0.00043	0.003	0.66
1.8		< 0.0003	0.00036	3.38
0.34		< 0.00003	0.000090	0.52
1.3		< 0.004	< 0.004	5.0
0.8				0.90
8.2				6.5
12.7		0.0049 ± 0.0007	0.35	1.7
130	149			46
8		< 0.002	0.007 ± 0.002	1.1
91	92	5.00 ± 0.04	5.89 ± 0.04	0.70
3.4		0.008 ± 0.001	0.016 ± 0.002	0.69
	125	2.70 ± 0.08	0.43 ± 0.04	130
		0.079 ± 0.008	0.28 ± 0.01	0.10
	3.6	0.86 ± 0.22	0.98 ± 0.76	
0.003		0.2 ± 0.03	0.39 ± 0.05	0.010
	11			
	133			

[a] Determined by Flameless Atomic Absorption Analysis.

residue, 16%; process naphtha, 5.3%–9.7%, and process solvent, −7.5%–7.5%.

Because of the variation of the mass flow under various operating conditions, the SRC mass balances are somewhat uncertain. The values used in our calculations were as follows: SRC, 60%; mineral residue,

Table VII. Trace and Minor Elements in

Element	Feed Coal		Solvent-Refined Coal (SRC)		Mineral Residue	
	NAA	XRF	NAA	XRF	NAA	XRF
Na	180		8.8		3150	
Fe	22,900	21,000	270	300	133,000	145,000
Cr	18	17	7.5	6.0	150	161
Co	5.2		0.26		32	
Ni	21	16	< 6	2.1	120	
Sc	2.6		0.45		14.7	< 110
Ba	46		0.14		240	
Sr	59		0.96		310	340
La	8.9		0.10		46.8	
Ce	17		0.4		94	
Pr	2.3		0.045		12	
Nd	9.3		0.27		44	
Sm	1.76		0.11		8.9	
Eu	0.36		0.025		1.76	
Gd	1.6		0.14		6.2	
Tb	0.26		0.026		1.05	
Ho	0.25		0.030		1.4	
Er	0.70		0.065		3.4	
Tm	0.12		0.016		0.53	
Yb	0.73		0.094		3.5	
Lu	0.11		0.014		0.50	
Hf	0.44		0.054		2.3	
Ta	0.14		0.043		0.62	
Th	1.9		0.19		10	
U	1.1		0.54		7.3	
Rb	13	15		< 0.4	77	97
Cs	0.95		0.012		5.8	
As	19	17	2.1	1.8	77	75
Sb	1.4		0.066		7.4	
Zn	26	25	8.1	7.2	120	138
Br	3.6	4.2	4.7	4.8	4.9	
Se	3.6		0.17		20	15
Hg[b]	0.16		0.02		< 0.005	
Ca		3		0.4		13
Cu		10		0.6		154
Pb		8		< 1		42

[a] Where estimated precision is not shown in this table, see Table VIII.

16%; process naphtha, 8%; and process solvent, $+2\%$. The process solvent and naphtha are not plotted in Figure 5 because in every case but Br, their contribution is less than 0.1%. For Br the contribution is 0.5%.

The major imbalance in both systems occurs for the volatile metal Hg. Substantial As is also unaccounted for in the SRC plant. These

Solvent-Refined Coal Materials in ppm[a]

Process Solvent	Process Naphtha	NBS Coal	
NAA	NAA	NAA	XRF
1.7 ± 0.03	0.51 ± 0.01	420	
14 ± 1 ppm	58 ± 2 ppm	8100	7800
0.044 ± 0.009	0.54 ± 0.02	19	
0.0026 ± 0.0005	0.0044 ± 0.0006	5.2	
		16	14
0.0007 ± 0.0001	0.0044 ± 0.0001	3.4	
< 1	< 2	390	
		170	150
< 0.001	0.0125 ± 0.008	10.5	
< 0.0002	0.0023 ± 0.0001	1.7	
< 0.0001	< 0.0001	0.28	
< 0.0002	< 0.0002	0.23	
< 0.003	< 0.004	0.97	
		0.46	
< 0.0009	< 0.001	3.4	
< 0.05	< 0.06	19	
< 0.001	< 0.002	1.4	
0.013 ± 0.001	0.105 ± 0.002	5.7	4.6
< 0.002	0.006 ± 0.001	3.7	
0.58 ± 0.04	0.86 ± 0.04	37	32
0.018 ± 0.002	0.25 ± 0.004	17	17
< 0.008	< 0.009	3.3	
0.07 ± 0.01	< 0.01	0.12	
			5
			15
			> 11

[b] Determined by Flameless Atomic Absorption Analysis.

imbalances are not surprising in view of the elevated temperature at which both of these processes operate. These metals are presumably lost in the offgas streams as is indicated in the offgas data in Table IV for the Laramie 10-ton retort, although it is possible that at least some of the Hg amalgamated with metal pipe and valve material in the plant. Gas stream measurements will be made in future SRC and oil shale retort

Table VIII. Estimated 1-σ Precision of Trace Element Analyses by Neutron Activation and X-Ray Fluorescence

Instrumental and Radiochemical Neutron Activation

1%	Na, Sc, Sm, Eu
3%	Cr, Co, La, Fe
5%	Nd, Gd, Tb, U, Th, As
10%	Pr, Ho, Hf, Sb, Yb
15%	Ce, Lu, Ta, Rb, Cs, Sr, Ba
20%	Er, Tm

X-Ray Fluorescence

5%	Sr, Fe
10%	Rb, As, Zn, Se
20%	Cr, Ni, Ga, Cu, Pb

runs in order to improve the material balances for As and Hg. Surprisingly, the volatile element Br gives a somewhat better balance in the SRC process than might be expected. This lack of expected volatility may indicate that Br is present largely in high molecular weight organic compounds or as the Br⁻ ion. Balances for all of the more refractory elements are relatively good. The apparent 16% imbalance for U probably is not significant.

Table IX. Trace and Minor Element Concentration Ratios Feed Coal/Solvent-Refined Coal

Element	Concentration	Element	Concentration
Na	20	Tm	4
Fe	85	Yb	8
Cr	2.5	Lu	4
Co	20	Hf	8
Ni	8	Ta	3
Sc	6	Th	10
Ba	104	U	2
Sr	74	Cs	79
La	73	As	9
Ce	42	Sb	21
Pr	51	Zn	3
Nd	34	Br	0.9
Sm	16	Se	21
Eu	15	Hg	8
Gd	5	Ga	8
Tb	10	Cu	17
Ho	8	Pb	> 8
Er	11		

Organic Results

Because the organic composition of these materials is very complex and the analytical techniques are still not resolved completely, it is impossible at present to give an organic characterization equivalent to the inorganic characterization shown in Tables V, VI, and VII. Our efforts

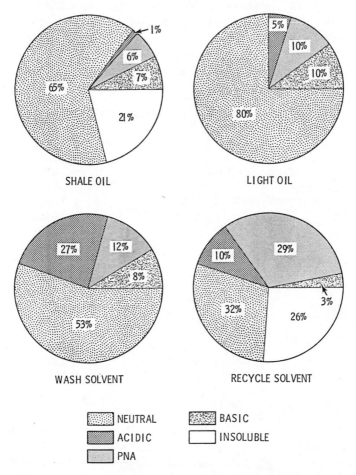

Figure 6. Proportions of shale oil and coal liquid fractions

have been directed toward developing a data base by using simple separation schemes with GC and GC–MS techniques. This data base then can be used to compare various materials and processes. We also have tried to use a flexible analytical methodology which can be adapted quickly to obtain specific information relevant to environmental or process related problems.

The characterizations are necessarily incomplete since only the major components were determined. Knowledge of the composition of a given fraction allows us to assess the need of further fractionation or to devise appropriate methods for the detection and measurement of minor components. This approach is essential in order to obtain useable information from bioassay screening tests.

Oil Shale Retort Offgases. Analysis of the retort offgas for organic constituents before the offgas burner showed mostly hydrocarbons. Saturated hydrocarbons, alkenes, alkynes, cyclic alkanes, and cyclic alkenes were found. Aromatic compounds, mostly alkyl-substituted benzenes, were detected. No heterocyclics were identified in the offgas.

Table X.   Partial Quantitative Analysis of Laramie Shale Oil (Run 13)

Naphthalene	1175 ± 50 ppm
2-Methylnaphthalene	700 ± 50 ppm
1-Methylnaphthalene	1150 ± 150 ppm
Fluorene	1025 ± 50 ppm
Anthracene–phenanthrene	575 ± 50 ppm
Fluoranthene	83 ± 12 ppm
Pyrene	140 ± 15 ppm
Quinoline	~ 500 ppm
Methylquinoline	~ 500 ppm
Carbazole	~ 1000 ppm
$C_{12}H_{26}$	9400 ± 1500 ppm
$C_{13}H_{28}$	8600 ± 510 ppm
$C_{14}H_{30}$	6150 ± 320 ppm
$C_{17}H_{36}$	8130 ± 990 ppm
$C_{18}H_{38}$	7170 ± 930 ppm
$C_{20}H_{42}$	3930 ± 350 ppm
$C_{21}H_{44}$	4700 ± 320 ppm
$C_{30}H_{62}$	1500 ± 480 ppm

Shale Oil and Solvent-Refined Coal Materials. The analysis of shale oil and coal liquids was carried out by application of the procedure outlined in Figure 3. The proportions of the resulting fractions obtained from our sample of Laramie shale oil (Run 13) are shown in the shale oil portion of Figure 6. A preliminary characterization is presented in Table X. It should be noted that the compounds listed are similar to those described in earlier work with other shale oils and other separation methods (12).

The principal components of the coal liquids were the neutral compounds and the aromatic compounds including PNA. (See Figure 6—light oil, wash solvent, and recycle solvent.) For the shale oil we analyzed, aliphatic hydrocarbons were most abundant. Lesser amounts of alicyclic

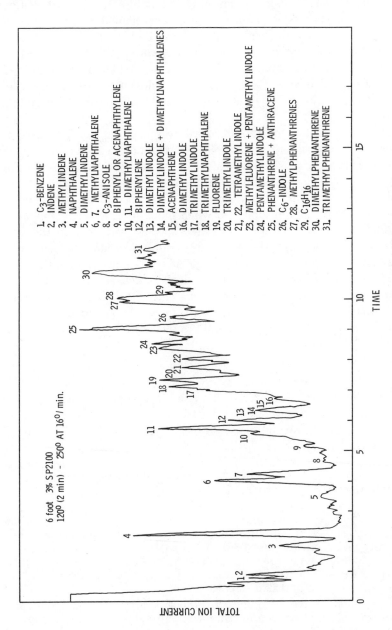

Figure 7. Typical gas-liquid chromatogram of shale oil PNA fraction

hydrocarbons were also present. Phenols are an important fraction (27%) of coal liquids but a minor fraction in shale oil (1%). The distribution of the phenolic compounds appeared to be very similar in both types of samples, being mostly alkyl-substituted phenols, although this result may be somewhat biased by our partitioning procedure. N-heterocyclic compounds were more abundant (≥ 7%) and in broader distribution and greater variety in shale oil than in the coal liquids. Hydroaromatics were important in the coal liquids but were negligible in the shale oil sample analyzed.

The PNA fraction of the shale oil was smaller (6%) than that fraction in the coal liquids (10–29%). In shale oil, a larger fraction of the PNA compounds are alkylated than in coal-derived liquids. For example, C_5 or higher-substituted aromatics were seen in shale oil but C_3 substitution was rare in coal liquid. This characteristic difference in alkyl substitution was repeated also when the N-heterocyclic compounds were similarly compared. Few alkylated species were seen in the coal liquids but C_6 and higher-substituted pyridines, quinolines, acridines, indoles, and carbazoles were detected in shale oil. For example, the PNA fraction of shale oil contained many indoles which can be seen in the gas chromatogram of this fraction (see Figure 7). The different alkyl substitution patterns found in these two syncrude materials may well reflect the underlying structural differences in coal and kerogen.

Conclusions

The following conclusions are based on the limited number of samples analyzed thus far (with due regard for variations that might be expected under different process conditions and feed materials).

1. Broad spectrum inorganic and organic analytical techniques provide the best approach for the initial characterization of the complex samples encountered in working with energy technologies such as oil shale retorting and solvent refining of coal.

2. Analytical methods characterized by good precision and sensitivity are necessary so that analytical errors will not add to other errors in difficult sampling situations, and so that the true magnitude of differences caused by process and feedstock variables can be measured. In complex samples, high precision inorganic analyses are facilitated by techniques such as neutron activation and x-ray fluorescence which are relatively insensitive to matrix effects.

3. A comparative organic constituent analysis of the particular crude shale oil and coal liquid samples analyzed in this study showed that the coal liquids contained higher concentrations of aromatic compounds including PNA's. The coal liquids were considerably richer in phenols than was the shale oil. N-heterocyclics were present in higher concentration in shale oil, corresponding to the higher nitrogen content

of the raw shale. Hydroaromatics were found to be common in coal liquids but negligible in this shale oil (*12*). We found measurable amounts of the heavy elements Hg, As, Zn, and Se in effluent streams from oil shale retorting. The process water also contained measurable amounts of Co, Br, and Sb. SRC liquids were relatively low in most trace elements. Most trace elements were concentrated in the mineral residue. Only Br was not depleted in SRC. Other trace elements remaining in measurable amounts were U, Ta (tantalum), Cr, and Zn. We have not yet measured the trace elements in gaseous and particulate samples from the SRC plant.

4. Satisfactory mass balances for the SRC pilot plant and the Livermore oil shale retort can be made for most trace elements measured to date. The major exception is the volatile element Hg, which will probably balance somewhat better after careful measurements of the offgas streams have been made.

Acknowledgments

The authors are indebted to R. W. Sanders, M. R. Smith, J. H. Reeves, C. L. Nelson, and M. R. Grove for their assistance with the analyses. E. S. Getchell, R. M. Garcia, and G. M. Garnant provided assistance with preparation of the manuscript. We are also indebted to the staffs of the Laramie Energy Research Center, Lawrence Livermore Laboratory's Oil Shale Project, the Pittsburg and Midway Coal Company, and Phyllis Fox of the University of California, Berkeley, for providing us with samples. This research was supported by ERDA Contracts RPLS-2126 and RPLS-1654.

Literature Cited

1. Gordon, G. E., et al., *Geochim. Cosmochim. Acta* (1968) **32**, 369.
2. Abel, K. H., Rancitelli, L. A., ADV. CHEM. SER. (1975) **141**, 118.
3. Soete, D. D., Gijbels, R., Hoste, J., "Neutron Activation Analysis," p. 234, Wiley, New York, 1972.
4. Laul, J. C., Schmitt, R. A., *Proc. Lunar Sci. Conf., 4th, 1973,* 1349.
5. Giauque, R. D., Goulding, F. S., Jaklevic, J. M., Pehl, R. H., *Anal. Chem.* (1973) **45**, 671.
6. Nielson, K. K., *Anal. Chem.* (1977) **49**, 641.
7. Braman, R. S., Foreback, C. C., *Science* (1972) **182**, 1247.
8. Braman, R. S., Johnson, D. L., *Environ. Sci. Technol.* (1974) **12**, 996.
9. Johnson, D. L., Braman, R. S., *Chemosphere* (1975) **4**, 333.
10. Fox, J. P., McLaughlin, R. D., Thomas, J. F., Poulson, R. E., *Colo. Sch. Mines, Proc. Oil Shale Symp., 10th,* John B. Reuben, Ed., p. 223, 1977.
11. Filby, R. H., Shah, K. R., Sautter, C. A., "Modern Trends in Neutron Activation Analysis Conference (1976)," *J. Radioanal. Chem.* (1977) **37**, 693.
12. Thorne, E. N., Ball, J. S., "The Chemistry of Petroleum Hydrocarbons," Reinhold, 1954.

RECEIVED October 17, 1977.

19

Chemical Class Fractionation of Fossil-Derived Materials for Biological Testing

BRUCE R. CLARK, C. H. HO, and A. RUSSELL JONES

Analytical Chemistry Division, Oak Ridge National Laboratory, Oak Ridge, TN 37830

Chemical fractionation of whole products and by-products from synthetic fuel production affords a logical first step in the evaluation of these materials for biological activity and the subsequent prediction of health hazards. Aliphatic and aromatic hydrocarbons, along with smaller amounts of heteroatomic species, constitute the bulk of all crude product materials and define a primary class separation need. Subfractionation of these fractions can lead to identification of bioactive components. Aliphatics are separated from the entire sample by a simple liquid chromatographic elution scheme. Aromatic compounds can be isolated by a cyclohexane–dimethylsulfoxide solvent partitioning scheme. A Sephadex LH–20 gel separation scheme appears feasible for the fractionation of crude liquids into aliphatic–aromatic, lipophilic–hydrophilic, polymeric, and hydrogen bonding classes of compounds.

There are many practical reasons to separate and analyze a complex, organic mixture according to chemical compound classes. One obvious example is the monitoring of a process stream for quality control with respect to a desirable or undesirable class of chemicals. For some purposes, a select group of compounds may be of interest, e.g., methyl-substituted, four-membered ring, polycyclic aromatic hydrocarbons (PAH). In such a case, a means for isolating the PAH class of compounds quantitatively and as directly as possible is needed. In this laboratory, class separations and analyses are integral parts of a multi-disciplined program to develop biological screening techniques in order to assess fossil-derived materials (products and by-products) for relative biological activities. This program is aimed toward developing some rapid and economical bioassay schemes which will reliably predict carcinogenic, mutagenic, teratogenic, and other physiological effects of exposure to potentially hazardous fossil-derived materials. In order to isolate components which foment biological activity, initial bioassays are performed with fractions of the whole sample. These fractions are produced most conveniently by chemical class fractionation schemes. Active fractions are consequently subfractionated and again bioassayed. Chemical analyses carried out interactively with bioassays provide a kind of feedback loop arrangement which can lead to the identification of a narrow class of compounds or even a specific compound responsible for a high rate of biological activity. This sort of iterative assay approach appears to be the most feasible for determining the presence of potentially hazardous substances in extremely complex and varied materials likely to be a part of a synthetic fuels industry and, consequently, a part of our environment.

Experience in class separations and analyses of fossil-derived materials began with the petroleum industry. The literature in this area is far too extensive to review here. Furthermore, petroleum literature deals principally with physical and chemical analyses of distillate fractions which are important to product characteristics. Recently, asphalts have received increased attention since they contain a wide range of known hazardous compounds. Most methods applied to whole samples of petroleum crudes have proven inadequate when dealing with synthetic coal liquids or shale oils because of stable emulsion formation in separation steps caused by larger amounts of inorganic and hydrophilic compounds.

Some specific class analyses have been reported for whole samples and distillates of shale oil and synthetic coal liquids (e.g., References 1, 2, 3, and 4). A few examples of specific class analyses recently reported include determination of hydrocarbon types in shale oil distillates by a hydroboration technique (5), aromatic–aliphatic group analyses in petroleum and coal liquids (6), nitrogen base types in high-boiling petroleum distillates (7), and PAH fractions in oils (8, 9).

Separating a whole sample of a coal liquid or shale oil into classes poses special problems since these materials contain high concentrations of heteroatomic species compared with natural petroleums. Many of these compounds are quite polar and can cause emulsification, precipitation, and may even react to produce artifactual compounds at some stage during a separation procedure. Many liquid chromatographic techniques have been useful in class separations and analyses of petroleums. More often, these have been applied to particular analytical scale operations with fossil-derived liquids. The most common applications are for aromatic–aliphatic and molecular weight types of separations.

In this paper we outline techniques used to isolate two specific classes of compounds as well as a general class separation method suitable for large scale production of fractions. The specific isolation schemes are devised to produce an alkane or a PAH-enriched fraction. The general scheme provides for rapid fractionation of whole samples into several broad chemical classes.

Alkane Class Isolation

Alkanes or saturated compounds constitute a class which is often a major fraction of synthetic crudes and always the major fraction in natural crudes. Many direct isolation methods have been devised for petroleum crudes and these are well documented in reviews (10). Synthetic crudes are handled very easily by dissolving a weighed portion in benzene, filtering, washing, removing the solvent, purifying on alumina, and analyzing by gas–liquid chromatography (GLC). Typically, a crude liquid is handled as follows: (1) ~ 1 g is dissolved in 200–300 mL of benzene, (2) the mixture is agitated and filtered, (3) 0.1N HCl (four 250-mL portions) followed by 0.1N NaOH aqueous solutions are used to wash the benzene solution, (4) the benzene is stripped on a rotary evaporator under reduced pressure at $\sim 40°C$, (5) the residue is dissolved in a few mL of hexane and placed on an alumina column (Activity I) where elution is carried out with hexane (about five bed volumes, i.e., 250 mL eluted $\sim 100\%$ of C_{11}–C_{32}, but 400 mL were used to assure complete recovery up to $\sim C_{40}$ or more), (6) the hexane is stripped in the same way as the benzene and the final residue is weighed, (7) hexane is added to the residue to make a solution for GLC analysis which is carried out on a 3% Dexsil 400 packed column.

Aqueous by-product samples from a coal liquefaction process, a coal gasification process, and a simulated in situ oil shale burn were extracted with successive 250-mL portions of benzene at neutral, acidic, and basic pH's. The benzene portions were combined, the solvent was stripped, and the remaining steps were carried out as described for the crude samples.

For all the samples, the peak areas of the chromatograms were integrated and compared with an external calibration mixture of n-alkanes. Identifications were confirmed from GLC-mass spectral data. Virtually no unsaturates were evident in the mixture, but a small amount of alkyl-substituted benzenes was apparent at short retention times. Quantitative data for two coal liquefaction crudes, a shale oil crude, a Louisiana–Mississippi sweet petroleum crude (LMS) and a mixed petroleum crude (Alaskan, Canadian, Californian, Iranian, Arabian, and LMS) are shown in Table I. These studies indicated that petroleum and shale crudes have

Table I. Determination of *n*-Alkanes in a Natural Petroleum, a Mixed Petroleum Crude, and Several Synthetic Crudes

Concentration (mg/g)

Compound	Mixed Petroleums	Crude Petroleum (LMS)	Shale Oil	Coal Crude A	Coal Crude B
C_{11}	6.11	5.31	7.55	0.34	0.12
C_{12}	6.06	5.84	6.65	4.41	0.44
C_{13}	6.31	6.87	6.20	2.98	0.45
C_{14}	6.18	6.75	4.80	1.30	0.03
C_{15}	4.92	5.46	3.30	1.85	0.39
C_{16}	4.65	5.54	4.50	3.14	1.08
C_{17}	3.86	5.14	8.60	2.51	0.45
C_{18}	6.03	6.41	8.80	3.64	1.48
C_{19}	3.95	4.68	5.60	1.82	0.47
C_{20}	3.83	4.39	4.80	1.85	0.41
C_{21}	3.63	4.38	5.70	2.40	0.96
C_{22}	3.52	4.17	4.50	1.73	0.46
C_{23}	3.51	4.43	5.00	2.60	0.69
C_{24}	2.73	3.27	3.60	1.98	0.34
C_{25}	2.19	2.65	3.90	1.99	0.32
C_{26}	1.95	2.30	3.45	2.09	0.31
C_{27}	1.94	2.26	3.35	2.22	0.34
C_{28}	1.28	1.63	2.90	1.77	0.31
C_{29}	1.06	1.48	3.20	1.55	0.30
C_{30}	0.89	1.10	1.00	0.96	0.23
C_{31}	0.86	1.08	0.20	0.86	0.22
C_{32}	0.94	1.02	0.10	0.86	0.20
C_{33}	0.43	0.73	0.65	0.24	0.12
C_{34}	0.47	0.62	0.40	0.23	0.13
C_{35}	0.38	0.51	—	0.17	0.10
C_{36}	0.22	0.33	0.15	0.11	0.09

high amounts of saturated components compared with either of the coal liquefaction crudes. This result is certainly expected from previous experience. A more noteworthy point is that the two coal liquids are quite different in kinds and amounts of *n*-alkanes. These data, of course, reflect the compositions of only these particular samples. It is understood generally that coal and shale liquid crudes can be hydro-treated to increase the degree of saturation to desired levels. Another matter which deserves comment is the absence of low molecular weight alkanes. It can be argued that these were lost in the rotary evaporation steps, but since the synthetic crudes were all produced under high temperature conditions and cooled at low pressures, most alkanes up to ~ C_{11} would be depleted in the samples.

Aqueous samples contained alkanes at the ppb level as might be expected. Shale oil retort water, produced in intimate contact with the

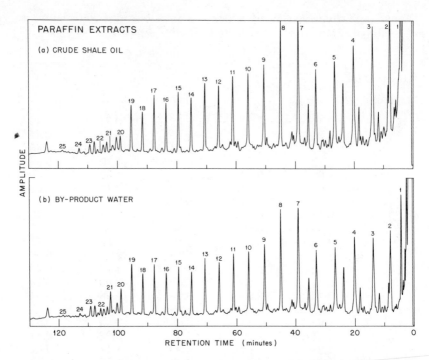

Figure 1. GLC chromatograms of the paraffin class isolates from crude shale oil and the intimately formed by-product water. GLC parameters: 1/8 in. × 10 ft glass column packed with 3% Dexsil 400 on Chromosorb W AW; temperature program 100°C (8 min); to 320°C (2°/min); 5-μl sample; flame ionization detection.

oil, showed the highest alkane content. Interestingly, the relative amounts of each alkane in the water paralleled almost exactly the relative amounts in the crude shale oil. This strongly suggests that the alkanes are not truly dissolved, but are likely held as surface coatings on unfiltered, finely divided solids which are evident in the water. The chromatograms in Figure 1 illustrate this point and also show the nearly pure aliphatic character of the isolates (the peaks at nonregular intervals are branched alkanes).

PAH Class Fractionation

No explanation or amplification of the importance of this class is necessary. We are interested especially in developing a systematic approach for obtaining PAH isolates in large quantities from a variety of fossil-derived liquids. The objectives are twofold: (1) to have available large, common stocks of these isolates for biological and environmental testing and research, and (2) to be able to provide, for any interested parties and for ourselves, PAH standard reference materials, i.e., isolates which have been characterized appreciably with respect to chemical content and stability.

Producing PAH isolates by a general scheme from complex, fossil-derived materials is made difficult by the wide variation in sample characteristics. The scheme presented here requires considerably more study to determine general applicability and actual separation efficiencies, but preliminary data indicate much promise in this approach. Figure 2 shows the sequence of steps which involve three-solvent distributions and two-column chromatographic purifications. A typical separation proceeds as follows: (1) 1-2 g of crude liquid are dissolved in 500 mL

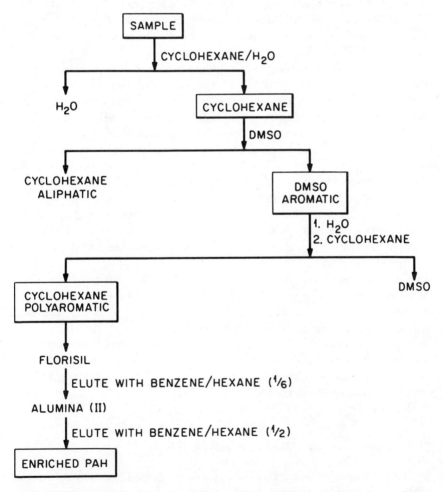

Figure 2. Separation scheme for polynuclear aromatic hydrocarbon subfractions from fossil-derived materials

of cyclohexane, (2) the mixture is washed by agitation while in contact with 500 mL of distilled water, (3) the mixture is left to settle until phase separation is complete, (4) the cyclohexane phase is withdrawn and filtered, (5) the cyclohexane phase is extracted with four 200-mL portions of dimethylsulfoxide (DMSO) which are recombined, (6) 1.2 L of distilled water is added to the DMSO (40% DMSO by volume), (7) four 500-mL portions of cyclohexane are used to back-extract the PAH material, (8) the combined cyclohexane portions are rotary evaporated to dryness, (9) the residue is dissolved in a small amount of hexane/benzene (6/1) and placed on a Florisil column and eluted with the same solvent mixture, (10) the volume of the eluate is reduced to a few mL and the solution is placed on an alumina column (Activity II), (11) the PAH isolate is eluted with the same hexane/benzene mixture and the final volume reduced to ~ 1 mL, and (12) the isolate is analyzed by GLC and GLC—MS on a 3% Dexsil 400 column.

The chromatograms indicate highly complex mixtures at this stage with evidence of some non-PAH components. But the addition of ^{14}C-labelled benz(a)anthracene or benzo(a)pyrene at the beginning of the entire procedure, indicates 95–98% recovery of these PAH compounds.

About 30 PAH compounds were identified tentatively by co-chromatographic methods and several of these identifications were confirmed from mass spectral data. Many components, however, could not be identified conclusively because of incomplete resolution from other components and the fact that isomers are not generally distinguishable by mass spectrometry. Clearly, the next step is to subfractionate the PAH isolate by techniques such as gel permeation or thin layer chromatography in order to facilitate making more positive identifications.

In these studies, the same fossil-derived crudes were used as described in the alkane fractionation work, i.e., two coal liquids, a shale oil, a petroleum, and a petroleum blend. Estimates of PAH components showed the highest overall amounts in shale oil and in one of the coal liquids. The petroleums contain the least amount and are comparable with the other coal liquid in this respect. Also, the petroleums have little detectable material of more than three rings, while the coal liquids contain appreciable quantities of five–six-ring PAH. Shale oil PAH compounds are predominantly in the 2–3-ring class with a high degree of alkyl substitution.

Class Separations with Sephadex LH–20 Gel

Gel permeation separations have been applied extensively to the analysis of organic mixtures, such as high molecular weight petroleum distillates and wood lignins. In 1966, Pharmacia (Sweden) introduced a gel, Sephadex LH–20, which could be used with polar solvents or with nonpolar organic solvents (*11*). Several specialized uses of this gel have appeared in the literature (*12, 13*); we have found the remarkable properties of this gel especially suitable to large scale separations with shale oil and two different coal-derived crude oils. When swollen and eluted with different solvents and solvent combinations, the gel behaves in dramatically different ways, extending its utility far beyond the normal permeation separation mode of behavior. The principal attractiveness of class separations using this gel centers around the gentleness of the method and the capacity of the gel to give excellent separations and virtually 100% recoveries with sample loads up to several hundred grams. Gentleness, i.e., no chemical or physical stresses, is especially important since nearly all usual separations use either heat, strong acids, strong bases, or ion exchange resins which almost assuredly alter the sample to some extent. Another advantage of LH–20 is its extreme stability; no degradation fragments have been detected.

The approach that we have developed thus far is outlined in Figure 3. The three boxes indicate the different modes of gel treatment which

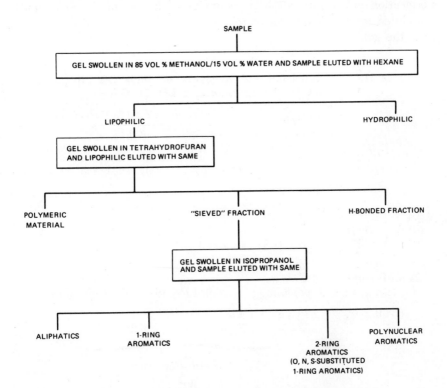

Figure 3. Separation scheme using Sephadex LH–20 gel

achieve the separations as outlined in the flow diagram. In the first instance, the gel is swollen with an 85/15 vol % of methanol/water. Treated in this manner, the gel is essentially a support for the polar methanol/water phase, and when hexane is used to elute a sample, the components are separated neatly into a lipophilic portion (eluted) and a hydrophilic portion (retained on the column). The hydrophilic fraction can be eluted with methanol or acetone. The lipophilic fraction is passed through a second column in which the gel has been swollen with tetrahydrofuran (THF) and is eluted with the same. Operated in this mode, the gel behaves primarily as a permeation gel. Polymeric lipophilic materials pass through with the void volume and are collected in the first fraction. A sieved fraction follows and, at high elution volumes, a small, distinct fraction is eluted. This last fraction is believed to be material that is retarded by hydrogen bonding mechanisms but was not separated into the hydrophilic fraction in the first step. In the last column, the combined sieved fraction is eluted through the gel with isopropyl alcohol with which the gel has also been swollen. Apparently, two separation mechanisms are operative here; the gel behaves as a permeation gel toward aliphatic compounds, but aromatic compounds are eluted later and in order of relative strengths of π-bond interactions with the gel.

Table II. Fractionation and Recovery of Shale Oil Lipophilic Fractions Using Sephadex LH–20 Gel

Shale Oil Sample Weight (g)	Lipophilic Fraction Recovery		Hydrophilic Fraction Recovery					
			with 85% MeOH		with Acetone		Total	
	(wt g)	(%)	(wt g)	(%)	(wt g)	(%)	(wt g)	(%)
17.5	16.2	92.6					1.3	7.4
17.9	17.1	95.1					1.1	6.1
94.6	88.2	93.2	5.2	5.5				
201.6 } 598.6	189.2	93.8	9.5	4.7 }	3.1	0.5	36.7	6.1
302.4	277.6	91.8	18.9	6.3				

Total Weight Partitioned	Total Recovered as Lipophilic	Total Recovered as Hydrophilic
634.0 g	588.3 g 92.8%	39.1 g 6.2%

Total Recovery

627.4 g
99.0%

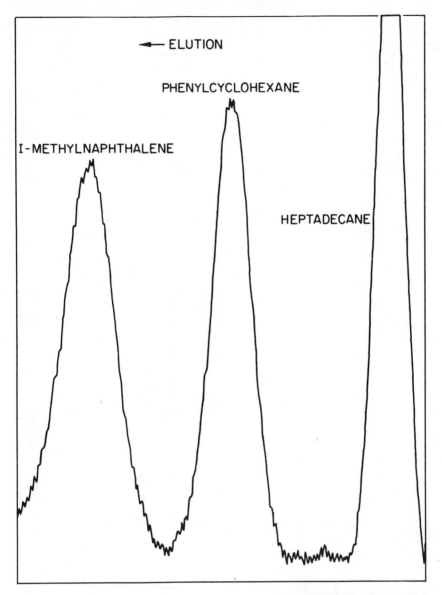

Figure 4. Demonstration of resolution between aliphatic and aromatic compounds in the third step of the separation scheme using Sephadex LH–20 gel

For lipophilic–hydrophilic separations the gel has a large sample capacity and the recoveries are quantitative as is demonstrated by the data summarized in Table II. A 5-cm × 1-m long column was used with solvent flow rates of ~ 500 mL per hr.

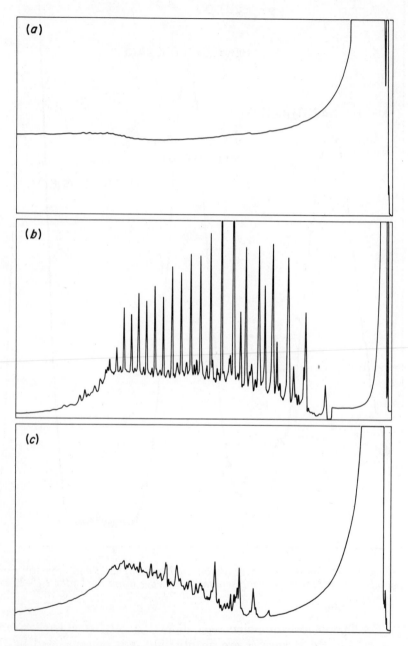

Figure 5.   GLC chromatograms of elution volumes corresponding to (a) a few void volumes, (b) the aliphatic fraction, and (c) the 2–3 ring aromatic fraction. GLC parameters same as in Figure 1.

Perhaps the most useful feature of LH–20 is its ability, when treated with isopropyl alcohol, to separate the aliphatic and aromatic hydrocarbons while at the same time eluting the aliphatic hydrocarbons in order of their decreasing molecular size and the aromatic hydrocarbons in order of their increasing aromatic ring content. No overlap of aliphatic–aromatic hydrocarbons occurs, even when the former have large alkyl substituents which decrease the elution volume of aromatics. This fact is well demonstrated by the data shown in Figure 4. Heptadecane, phenylcyclohexane, and 1-methyl naphthalene are separated completely, showing the resolution between one- and two-ring aromatics and a low molecular weight aliphatic hydrocarbon. Substituent groups containing oxygen, nitrogen, and sulfur increase the elution volumes in both the aliphatic and aromatic series. Of course, most materials of this class should be concentrated in the hydrophilic and hydrogen-bonded fractions isolated in previous steps.

Preliminary examination of shale oil components in the eluates of the isopropyl alcohol elution step indicate behavior very similar to that observed with the model compound mixtures. Figure 5 shows chromatograms obtained from GLC analyses on a 3% Dexsil 400 column. The top chromatogram is the fraction obtained after elution of a few void volumes; no significant amounts of material are detected even at high sensitivity. The middle chromatogram is of the fraction having an elution volume which is sufficient to elute aliphatic hydrocarbons down to C_{10}; GLC–MS data indicate an entirely aliphatic series of components. The GLC profile of the first portion of eluate into the aromatic elution volume range is shown in the bottom chromatograph. Although only partial identifications have been made on this and succeeding fractions, GLC–MS data have confirmed that there are no aliphatic hydrocarbons present and that the major peaks are mostly alkyl-substituted, 2–3 ring aromatic hydrocarbons.

These studies all indicate that class fractionation with Sephadex LH–20 gel has a high potential for affording one of the most gentle, reproducible and efficient separation schemes yet reported for complex organic mixtures. There is a high probability that a PAH isolate can be produced more easily and in higher purity by this method than by the method outlined previously in this paper. The large sample capacities of the columns, commensurate with good separations, make this approach ideal for the preparation of samples for biological and environmental research. A more detailed and conclusive report on this work is found in Reference *14*.

Acknowledgments

We thank the Laramie and Pittsburgh Energy Research Centers for providing samples and for their continued collaboration in these studies.

Also, we hasten to point out that none of the product crudes or aqueous by-products studied here are purported to necessarily represent any product or by-product from a likely, commercial scale process. These samples are deemed to provide reasonable representations for the purpose of methods development.

This research was sponsored jointly by the Environmental Protection Agency and the U.S. Energy Research and Development Administration under contract with the Union Carbide Corporation.

Literature Cited

1. Dooley, J. E., Sturm, G. P., Jr., Woodward, P. W., Vogh, J. W., Thompson, C. J., "Analyzing Syncrude from Utah Coal," Bartlesville Energy Research Center, BERC RI–75/7, 1975.
2. Sturm, G. P., Jr., Woodward, P. W., Vogh, J. W., Holmes, S. A., Dooley, J. E., "Analyzing Syncrude from Western Kentucky Coal," Bartlesville Energy Research Center, BERC RI–75/12, 1975.
3. Woodward, P. W., Sturm, G. P., Jr., Vogh, J. W., Holmes, S. A., Dooley, J. E., "Compositional Analyses of Synthoil from West Virginia Coal," Bartlesville Energy Research Center, BERC RI–76/2, 1976.
4. Poulson, R. E., Frost, C. M., Jensen, H. B., "Characteristics of Synthetic Crude Oil Produced by In Situ Combustion Retorting," *Am. Chem. Soc., Div. Fuel Chem., Prepr.* (1974) 19(2), 175–182.
5. Jackson, L. P., Allbright, C. S., Jensen, H. B., "Semimicro Procedure for the Determination of Hydrocarbon Types in Shale Oil Distillates," *Anal. Chem.* (1974) 46, 604.
6. Suatoni, J. C., Swab, R. E., "Rapid Hydrocarbon Group-Type Analysis by High Performance Liquid Chromatography," *J. Chromatogr. Sci.* (1975) 13, 361.
7. McKay, J. F., Weber, J. H., Latham, D. R., "Characterization of Nitrogen Bases in High-Boiling Petroleum Distillates," *Anal. Chem.* (1976) 48, 891.
8. Lee, M. L., Bartle, K. D., Novotny, M. V., "Profiles of the Polynuclear Aromatic Fraction from Engine Oils Obtained by Capillary Column Gas–Liquid Chromatography and Nitrogen-Selective Detection," *Anal. Chem.* (1975) 47, 540.
9. Popl, M., Stejskal, M., Mostechy, J., "Determination of Polycyclic Aromatic Hydrocarbons in White Petroleum Products," *Anal. Chem.* (1975) 47, 1947.
10. Trusell, F. C., "Crude Oils–Hydrocarbons," *Anal. Chem.* (1975) 5, 169R.
11. Joustra, M., Soderquist, B., Fischer, T.-L., "Gel Filtration in Organic Solvents," *J. Chromatogr.* (1967) 28, 21.
12. Marin–Mudrovcic, S., Muhl, J., *Sateva, M., Nafta (Zagreb)* (1972) 23, 593.
13. Klimish, H. J., Ambrosius, D., "Gel Chromatography of Polycyclic Aromatic Hydrocarbons," *J. Chromatogr.* (1974) 94(1), 311.
14. Jones, A. R., Guerin, M. R., Clark, B. R., "Preparative-Scale Liquid Chromatographic Fractionation of Crude Oils Derived from Coal and Shale," *Anal. Chem.* (1977) 49, 1766.

RECEIVED August 5, 1977.

HPLC Techniques for Analysis of Residual Fractions

HARRY V. DRUSHEL

Exxon Research and Development Laboratories, Exxon Company, U.S.A.,
P. O. Box 2226, Baton Rouge, LA 70821

*The development of processes for conversion of petroleum
residua and tar sands bitumen to useful products requires
analytical techniques to follow changes in composition.
High-pressure liquid chromatographic (HLPC) techniques
lend themselves well to such studies because they provide
information regarding composition with a minimum of time
and effort compared with conventional large scale chromatographic separations. The use of μ-Porasil columns with
backflush capability was investigated for separation of saturates and aromatics. The use of differential refractive index
and moving-wire detectors were compared for use in quantitation of the saturates and aromatics. Factors affecting
their response were studied and the results were compared
with compositions determined by the large scale clay–silica
gel separation technique.*

The upgrading of petroleum residua, tar sands bitumen, etc., is
becoming increasingly important as petroleum reserves diminish.
The development of processes for their conversion to useful products
requires analytical techniques to follow changes in composition. High-
pressure liquid chromatographic (HPLC) techniques lend themselves
well to such studies because they provide information regarding composition with a minimum of time and effort compared with conventional
large scale chromatographic separations.

Some of the more complicated separation–characterization schemes
used to analyze high-boiling residua and related materials by Drushel
(1, 2) and Jewell and co-workers (3, 4) make use of large columns
packed with alumina or clay plus silica gel. These techniques may

require a lapsed time of up to two days to obtain quantitative data regarding the distribution of saturates and aromatics. One problem with such separations is the carry-over of aromatic hydrocarbons into the saturates fraction if UV monitoring is not used to determine the proper cut-off in the separation. The retention of solvent in the fraction or the evaporation of the lower-boiling portion of the sample can cause error in the gravimetric determination of the concentration of saturates and aromatics.

This chapter describes the application of HPLC techniques for the determination of saturates and aromatics in high-boiling fractions. This study involves the use of a microparticulate silica column (μ-Porasil) with backflush operation to obtain the aromatics similar to the technique described by Suatoni and Swab (5, 6). The possibility of using a moving-wire detector which detects organic carbon in conjunction with the conventional differential refractive index detector was investigated. Variables affecting the response of these dual detectors were investigated and the results were compared with data from the more conventional clay–gel chromatographic separation.

Experimental

Liquid Chromatographic System. The basic equipment used in this study was a Waters Associates Model ALC 202/401 liquid chromatograph equipped with a Model 6000 pump and a Waters Model R–401 differential refractive index (RI) detector. The UV detector included with this equipment was not used in this study. A single μ-Porasil column was used. For most of the work, n-heptane was used as the solvent at a flow rate of 1 mL/min.

Moving-Wire Detector. A Pye Unicam LCM–2 moving-wire detector was used in series with the refractive index detector. The exit of the RI detector was connected to the moving-wire detector by means of a 4-ft length of .040-in. I.D. stainless steel tubing. Minimum volume stainless steel tubing could not be used because of the creation of excessive back pressure which could damage the delicate cell in the RI detector. The larger volume stainless steel tubing caused some lag in response and some mixing which resulted in broadening of the peaks. As a compromise, 0.020-in. tubing might perform better.

Sample Injection and Backflush Valves. For injection of the samples, a Valco 6-port valve (catalog no. CV–6–UHPA–N–60) with a 10-μL sample loop was used for sample injections. A high-pressure Valco 4-port valve (catalog no. CV–4–HPax–N–60) was used to switch the solvent flow through the columns to obtain backflush of the aromatic components after the appearance of the saturates peak. The overall arrangement of the chromatographic equipment showing the configuration of these valves is presented diagrammatically in Figure 1.

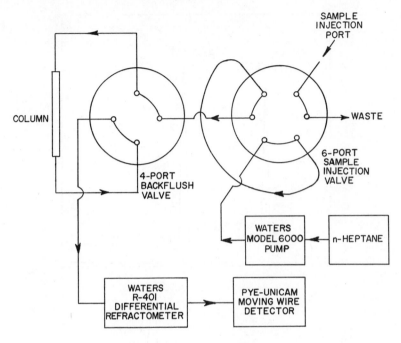

Figure 1.   Schematic of HPLC system

Discussion of Results

Typical Response Peaks. Figure 2 shows some typical response peaks for the HPLC separation of the saturates and aromatics in a typical vacuum gas oil. The curve in the upper portion of the figure is the response for the sample using the RI detector. After the saturates peak appeared, the backflush valve was switched and the attenuation changed as indicated in the figure. The peaks were very sharp and symmetrical and appeared in less than 10-min lapsed time from the point of injection. Base line drift was minimal and the areas of the response peaks were obtained with a ball and disc integrator on the strip chart recorder.

The response peaks for the saturates and aromatics using the moving-wire detector are shown in the bottom half of Figure 2. The response peaks were retarded by ~ 2 mL from the retention volume obtained with the RI detector. In addition, the peaks were broadened appreciably and the valley between the saturates and aromatics peaks did not return completely to the base line. Since there was a small tail on the aromatics peak as well, it was concluded that there was some hold-up in the system, most probably at the sampling point on the LCM–2 or in the 0.040-in. I.D. tubing.

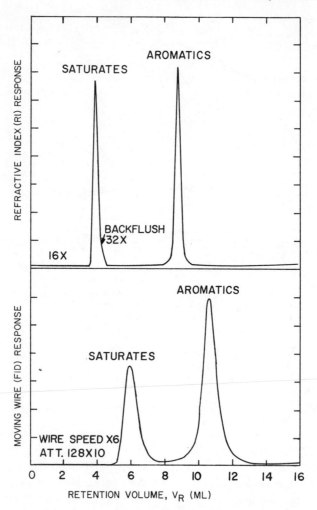

Figure 2. Typical response peaks for HPLC separation of saturates and aromatics. Column, 1-ft μ-Porasil; solvent, n-heptane (1 mL/min); sample, 10 μL of vacuum gas oil in n-heptane; detectors, refractive index, and Pye Unicam moving wire.

Effect of Sample Size on Response. The response of the dual detectors was checked for linearity by analyzing the vacuum gas oil sample at various dilutions. Results are presented in Table I. The data indicate that the response for both detectors is essentially linear over a 25-fold range in concentration. Over this range, the ratio of the response for the aromatics with respect to that for the saturates is constant for both detectors to within several percent average deviation. Also, these data show that the aromatics/saturates response ratio is greater for the RI detector than it is for the moving-wire detector under the conditions used.

Effect of Wire Speed on Relative Response. Response for the saturates and aromatics in the vacuum gas oil using the moving-wire detector was checked at each of the five wire speeds available from X1 through X6. If the square root of the response for the two peaks was plotted against the wire speed, a nearly linear plot for both the aromatics and the saturates was obtained as shown in Figure 3. In addition to this nonlinear response, the aromatics/saturates response ratio varied from just under 2 at a wire speed of X6 to a value near 4.6 at a wire speed of X1. A plot of the response ratio as a function of wire speed is shown in Figure 4. Although the response ratio approaches a limiting value at the highest speed of X6, a speed necessary to obtain the true response ratio is probably greater than can be achieved on the moving-wire detector as supplied by the manufacturer. Deviations in the response

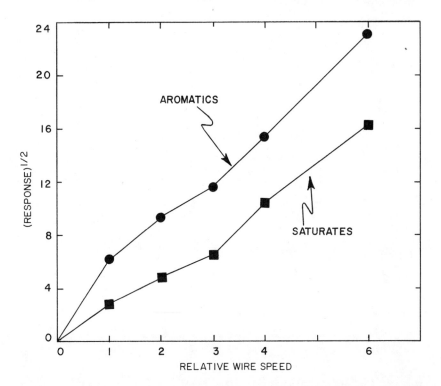

Figure 3.   Effect of wire speed on relative response (Pye Unicam detector). Sample, vacuum gas oil; solvent, n-heptane (1 mL/min); concentration, neat; sample volume, 10 μL; column, 1-ft μ-Porasil; mode, backflush operation for aromatics.

Table I. Variation of Response with Sample

Moving-Wire FID Response[b]

Sample Size (mg)	Attenuation	Area ($\times 10^{-3}$)		Ratio Aromatics/ Saturates
		Saturates	Aromatics	
8.96	1024 × 1	369.2 (41.2)	462.9 (51.7)	1.25
4.48	512 × 1	153.1 (34.2)	220.5 (49.2)	1.44
1.79	256 × 1	72.9 (40.7)	100.9 (56.4)	1.38
0.896	128 × 1	39.8 (44.4)	55.7 (62.2)	1.40
0.358	64 × 1	17.8 (49.7)	25.5 (71.2)	1.43

[a] Sample, vacuum gas oil; solvent, n-heptane (1 mL/min); sample volume, 100-μL loop (dilutions used to reduce sample size); column, 1-ft μ-Porasil; mode, backflush operation for aromatics.

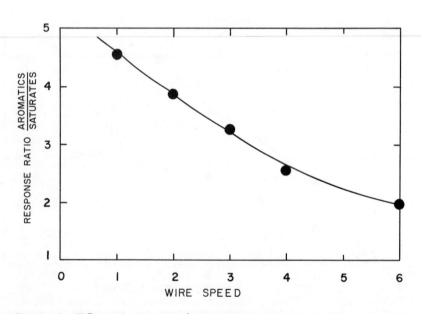

Figure 4. Effect of wire speed on aromatics/saturates response ratio. Sample, 10 μL of vacuum gas oil; solvent, n-heptane (1 mL/min); column, 1-ft μ-Porasil, mode, backflush operation for aromatics; detector, Pye Unicam moving-wire detector.

Size (Pye Unicam and RI Detectors)[a]

RI Response[b]

	Area (×10⁻³)		Ratio Aromatics/
Attenuation	Saturates	Aromatics	Saturates
128 NL[c]	28.3	41.5	1.46
	(3.16)	(4.63)	
128 NL[c]	15.4	27.5	1.79
	(3.44)	(6.14)	
64	6.13	11.97	1.95
	(3.42)	(6.69)	
32	3.16	5.94	1.88
	(3.53)	(6.63)	
16	1.22	2.38	1.95
	(3.41)	(6.65)	

[b] Evaporator temp., 200°C, oxidizer temp., 800°C, and wire speed ×6. Area response per mg of sample shown in parenthesis.
[c] Nonlinear range of the detector.

for the saturates and aromatics as a function of wire speed are probably reflections of the extent to which the solvent is retained by the solute deposited on the wire or to the rate at which lower-boiling components of the deposited solute are lost through vaporization.

Determination of Response Factors. Specific response factors were obtained by injecting known concentrations of saturates and aromatics fractions obtained by clay–gel chromatographic separation of various gas oil fractions as well as residua boiling above 510°C. All of the clay–gel saturates fractions showed the presence of some aromatic impurities (2–20%) by HPLC. This was particularly true of the saturates obtained from the 510°C⁺ residue samples. Also, the aromatics fractions showed the presence of some saturates (2–3%) by HPLC. The response factors for these saturates and aromatics fractions are listed in Table II. Based on the values shown in Table II, the response for the aromatics was about 1.7 times that for the saturates. The ratio of the response factors for the gas oil fractions differs from the ratio for the residuum samples by about 6%, relative.

Comparison of the HPLC Technique with Clay–Gel Chromatographic Separation. A number of vacuum gas oils were analyzed by preparing solutions in *n*-heptane at a concentration near 100 mg/mL. These samples were injected into the HPLC equipment and the concentration of saturates and aromatics calculated from the response factors shown in Table II. The absolute percentages of saturates and aromatics are shown in Table III along with the polar aromatics obtained by subtracting the sum of these from 100%.

Table II.   Response Factors Obtained

Sample	Distillate Cut[b]	Clay–Gel Cut	Concn. (mg/mL)
601	502	Saturates	55.79
816	502	Saturates	52.01
817	502	Saturates	50.31
815	502	Aromatics	65.29
816	502	Aromatics	56.04
548	502	Aromatics	59.81
687	503	Saturates	60.78
688	503	Saturates	59.05
689	503	Saturates	60.63
686	503	Aromatics	60.64
687	501	Aromatics	63.82
688	503	Aromatics	58.03

[a] Column and detector, 1-ft μ-Porasil, RI detector; solvent, n-heptane (1 mL/min); injector, Valco 6-port valve with 10-μL sample loop; mode, backflush operation for aromatics.
[b] Boiling range of cuts 502 is 340–510°C. Boiling range of cuts 503 is 510°C+ (distillation residue).

Table III.   Comparison of HPLC with

				Calculated Distribution (wt %) HPLC Method		
Sample	Cut[a]	Sulfur (Wt %)	Concn. (mg/mL)	Satu-rates	Aro-matics	Polars (100–Sum)
243	502	0.667	127.1	53.7	42.1	4.2
244	502	0.464	106.9	54.7	36.9	8.4
245	502	0.453	119.4	52.5	36.0	11.5
249	502	0.518	116.4	51.2	36.9	11.9
250	502	0.553	110.3	49.0	35.1	15.8
296	502	0.901	134.5	51.2	41.9	6.9

[a] Boiling range 340–510°C.

from Clay–Gel Chromatographic Fractions[a]

Area (integrator counts)		Relative Response Factor[c]	
Saturates	*Aromatics*	*K_S Saturates*	*K_A Aromatics*
2,880	22.4	51.8	—
2,650	92.8	51.8	—
2,528	226.4	52.5	—
73.6	5,824	—	91.5
118.4	4,736	—	88.7
25.6	5,472	—	92.3
	Averages:	52.0	90.8
2,576	1,232	52.5	—
2,640	1,048	53.6	—
2,688	760	50.6	—
57.6	5,248	—	88.4
64.0	4,928	—	84.0
64.0	5,344	—	85.8
64.0	5,344	—	85.8
96.0	4,960	—	88.8
64.0	4,896	—	86.6
64.0	4,864	—	86.0
64.0	4,864	—	86.0
	Averages:	52.2	86.4

[c] Response (in integrator counts) divided by concentration in mg/mL. Saturates were corrected for aromatic impurity and aromatics were corrected for saturate impurity.

Clay–Gel Chromatographic Separation

Normalized Distribution (wt %)					
HPLC Method			Clay–Gel Method		
Saturates	*Aromatics*	*Polars*[b]	*Saturates*	*Aromatics*	*Polars*
55.2	43.3	1.5	56.3	42.1	1.6
58.8	39.7	1.5	59.6	39.0	1.4
58.4	40.1	1.5	59.0	39.5	1.5
57.2	41.3	1.5	57.6	41.0	1.4
57.4	41.1	1.5	57.5	41.0	1.5
54.2	44.3	1.5	54.5	43.8	1.7
				Average:	1.5

[b] Assuming polars = 1.5 wt % for all samples.

Table IV. Repeatability of Saturates–Aromatics

Refractive Index Detector

Run	Area ($\times 10^{-3}$)		Area Ratio Arom./Sats.	Retention Vol (mL)	
	Saturates	Aromatics		Saturate	Aromatics
1	2.21	4.99	2.26	3.9	8.2
2	2.30	5.15	2.24	3.8	8.2
3	2.37	5.44	2.30	3.8	8.2
4	2.27	5.38	2.37	3.8	8.3
5	2.14	5.41	2.53	3.8	8.2
6	2.27	5.25	2.31	3.8	8.2
7	2.18	5.12	2.35	3.9	8.2
8	2.27	5.31	2.34	3.9	8.3
9	2.27	5.41	2.38	3.9	8.3
10	2.37	5.50	2.32	3.9	8.3
Average:	2.27	5.30	2.34	3.85	8.24
2σ:	±0.148	±0.329	±0.161	±0.105	±0.103
	($\pm6.5\%$)	($\pm6.2\%$)	($\pm6.9\%$)	($\pm2.7\%$)	($\pm1.3\%$)

[a] Sample, 10 μL of a 20-vol % solution of vacuum gas oil; solvent, n-heptane (1 mL/min); column, 1-ft μ-Porasil; mode, backflush operation for aromatics; injector, Valco 6-port valve with 10-μL sample loop.

Because of an apparent variation in the response for both saturates and aromatics from sample to sample, depending upon their origin, the results do not agree with the composition obtained by the clay–gel separation. In addition, most of the error is thrown into the polar aromatic category because it is calculated by difference and includes errors from both measurements. Since the percentage of polars amounted to only 1.5%, on the average, as determined by the clay–gel method, much better results were obtained by normalizing the results. In fact, the data presented for the clay–gel method were also normalized to correct for errors introduced by vaporization losses, column hang-up, and retained solvent. By using the normalization technique, the results for the HPLC method agreed very well with the clay–gel method. In fact, the results obtained by the normalized HPLC method agreed very well with data expected on the basis of the extent to which these samples had been hydro-treated.

Polar compounds containing oxygen and nitrogen atoms are strongly adsorbed and do not elute from a μ-Porasil column with n-heptane. Accumulated polar material could alter the resolution and capacity of the column after extended use. If a guard column is not used, regeneration techniques would be required to rejuvenate the column. Because of variations in response, accurate analyses can be obtained only by calibrating with saturates and aromatics fractions separated from the sample by a large scale chromatographic technique.

Separation (Pye Unicam and RI Detectors) [a]

Moving-Wire Detector

Area ($\times 10^{-3}$)		Area Ratio	Retention Vol (mL)	
Saturates	Aromatics	Arom./Sats.	Saturates	Aromatics
58.9	105.0	1.78	6.1	10.7
57.3	112.6	1.97	6.0	10.6
58.2	119.4	2.05	6.0	10.6
60.2	121.3	2.01	6.1	10.8
55.7	112.3	2.02	6.1	10.7
59.2	113.9	1.92	6.1	10.7
58.2	115.2	1.98	6.1	10.6
60.5	118.4	1.96	6.1	10.7
60.2	113.6	1.89	6.1	10.8
59.5	121.0	2.03	6.1	10.7
58.8	115.3	1.96	6.08	10.69
± 2.99	± 9.94	± 0.161	± 0.084	± 0.148
($\pm 5.1\%$)	($\pm 8.6\%$)	($\pm 8.2\%$)	($\pm 1.4\%$)	($\pm 1.4\%$)

Repeatability of the Saturates–Aromatics Separation. Results obtained with the dual detectors for ten individual injections of a 20-vol % solution of a vacuum gas oil in *n*-heptane are shown in Table IV. The precision (2σ) for the saturates and aromatics area response from the RI detector was $\sim \pm 6\%$, relative. The precision for the aromatics/saturates response ratio was 6.9%, relative. The retention volumes were much more reproducible and had 2σ values of the order of 1–3%, relative.

Table V. Comparison of Moving-Wire Detector with the Refractive Index Detector and Clay–Gel Chromatographic Separation (Vacuum Gas Oil)

Normalized Distribution (Wt %)

	Trapped from μ-Porasil	Clay–Gel Separation	Refractive Index Detector	Moving-Wire Detector
Saturates	38.2	42.6	40.5 [b]	31.8 [d]
Aromatics	56.6	52.2	54.3 [c]	63.0 [e]
Polars	5.2 [a]	5.2	5.2 [a]	5.2 [a]

[a] Using wt % polars as determined by clay–gel separation.
[b] Using factor $K_S = 52.0$ mL–ct/mg.
[c] Using factor $K_A = 90.8$ mL–ct/mg.
[d] Corrected for % carbon, assumed to be 86%.
[e] Corrected for % carbon, assumed to be 85%.

For the moving-wire detector, the precision (2σ) for the saturates response was $\pm 5\%$, relative, while for the aromatics it was $\pm 8.6\%$, relative. The aromatic/saturate response ratio was $\pm 8\%$, relative. The precision for the retention volume was $\pm 1.4\%$, relative.

Comparison of Moving-Wire and RI Detectors with Clay–Gel Chromatographic Separation. The response areas for the saturates and aromatics in the vacuum gas oil, as obtained for ten sample injections on the two detectors, are listed in Table IV. The average response areas were used to calculate the percentages of saturates and aromatics in the sample, which are shown in Table V. Agreement between the results obtained with the RI detector and those obtained by the clay–gel separation technique or by repetitive trapping from the μ-Porisil column itself is good. For the HPLC calculations it was necessary to assume the concentration of polars determined previously by the clay–gel separation technique. Results obtained by the moving wire did not agree as well with the clay–gel data. The saturates are low by 11%, absolute, and the aromatics are high by the same amount. Presumably, these differences were obtained because the proper operating parameters on the moving-wire detector were not achieved. Future studies will be directed toward studying these parameters to improve the accuracy of the moving-wire detector. This would permit its use in place of the RI detector so that solvent programming can be used for more complex HPLC separations.

Acknowledgment

The author thanks Exxon Research and Development Laboratories for permission to publish the results of this study.

Literature Cited

1. Drushel, H. V., *Am. Chem. Soc., Div. Pet. Chem., Prepr.* (1972) **17**(4), F92.
2. Drushel, H. V., *Am. Chem. Soc., Div. Pet. Chem., Prepr.* (1976) **21**(1), 146.
3. Jewell, D. M., Ruberto, R. G., Davis, B. E., *Am. Chem. Soc., Div. Pet. Chem., Prepr.* (1972) **17**(1), A55.
4. Jewell, D. M., Albaugh, E. W., Davis, B. E., Ruberto, R. G., *Am. Chem. Soc., Div. Pet. Chem., Prepr.* (1972) **17**(4), F81.
5. Suatoni, J. C., Swab, R. E., *J. Chromatogr. Sci.* (1975) **13**, 361.
6. Suatoni, J. C., Swab, R. E., *J. Chromatogr. Sci.* (1976) **14**, 535.

RECEIVED August 5, 1977.

Analytical Characterization of Solvent-Refined Coal Comparison with Petroleum Residua

R. B. CALLEN, C. A. SIMPSON, and J. G. BENDORAITIS

Mobil Research & Development Corporation, Paulsboro, NJ 08066

Gradient elution chromatography has been used to separate solvent-refined coals produced from three different coal sources into well defined fractions. These fractions then were examined by IR spectrometry to determine the nature of the functional groups, and by a combination of C-13 and H-NMR techniques to gain some insight into the aromaticities of the fractions. Particular attention has been paid to the distribution of heteroatoms N, O, and S within the chromatographic cuts. Data from these SRC's are presented and compared with comparable data from a petroleum vacuum resid.

Over the past several years there has been a renewed interest in coal liquefaction brought about principally by a desire to utilize the large coal reserves present within the continental United States. Several coal liquefaction pilot plants have been constructed or are in various stages of design, and it is anticipated that coal liquids will become commercially available during the 1980's. A number of different processes have been developed to produce clean liquid fuels from coal (1). One such method is the solvent-refined coal (SRC) process which produces a low-sulfur, low-ash fuel that can be used by utilities for power generation.

Although SRC's are lower in both sulfur and mineral matter than the coals from which they are derived, these materials will require additional upgrading if higher quality fuels, such as gas turbine fuels, are desired in the product slate (2). One attractive route for upgrading SRC is an extension of petroleum resid hydroprocessing technology (3). In order to fully understand the chemistry involved in hydroprocessing

0-8412-0395-4/78/33-170-307$05.00/1

SRC, a detailed examination of the composition of these coal liquids is of fundamental importance. Numerous procedures have been published previously for investigating the composition of liquids derived from coal. In general, these procedures combine separation techniques with a variety of spectroscopic methods to provide the desired quantity of structural information. The separation techniques used include methods based on solubility fractionation (4,5), methods combining solubility fractionation and adsorption chromatography (6), and liquid chromatographic procedures for chemical fractionation (7,8). Chemical reactions also have been used to separate coal liquid asphaltenes into acidic and basic fractions (9).

In this study, gradient elution chromatography (GEC) has been used to separate SRC's, produced from three different coal sources, into fractions that differ according to the polarity of the solute species. This particular GEC method provides sufficient separation of resins, asphaltenes, and highly functional polar asphaltenes while maintaining the ability to resolve the hydrogen-rich, coal liquid components into a saturate fraction and two distinct aromatic oil fractions. Thus, this separation scheme is quite appropriate for studying the chemical transformations that occur by hydroprocessing SRC's in which the production of aromatic oils occurs as heteroatoms are removed from the more polar fractions (3). The GEC fractions have been examined by C-13 and H-NMR techniques to gain some insight into the aromatic nature and structural features of the various fractions, and by IR spectrometry to determine the nature of functional groups present therein. Particular attention has been paid to the distribution of heteroatoms nitrogen, oxygen, and sulfur within the chromatographic cuts. Data from these SRC's are compared and contrasted with comparable data from a petroleum vacuum resid.

Materials

The SRC's used in this study were obtained from the SRC process demonstration unit operated by Southern Services Inc. in Wilsonville, Alabama. Three SRC's were studied and these products were derived from the following coals: Illinois No. 6 Burning Star, Illinois No. 6 Monterey, and Wyodak (Amax). The as-received SRC's were wet owing to a water quench and therefore each SRC was dried carefully before sampling and analysis. The vacuum resid was prepared by laboratory vacuum distillation of an atmospheric resid obtained from a commercial refinery source. The identification and source of the coal liquid samples used in this study along with other pertinent information are presented in Table I.

Table I. Identification and Source of SRC

Product

	Burning Star SRC	Monterey SRC	Wyodak Amax
Source	←——— SRC Pilot Plant in Wilsonville, AL ———→		
Run number	14	52	57C
Sample number	4787	10852	11423
Coal source	Illinois No. 6 Burning Star Mine	Illinois No. 6 Monterey Mine	Amax
Mobil sample No.	75D-42	75D-3018	75D-3326
P (psig)	1500	2444	2509
T (°F)	827	853	843
Conversion % of MAF feed	93	94.9	83.4

Methods of Analysis

All samples were analyzed directly for carbon, hydrogen, oxygen, nitrogen, and sulfur. Carbon and hydrogen were determined by Pregl-type microcombustion; oxygen was determined by the Unterzaucher method; nitrogen was determined by an automated micro-Dumas procedure, and sulfur was determined by ASTM Method D-129. Molecular weights were determined by vapor pressure osmometry (VPO) in either benzene or tetrahydrofuran (THF) depending upon solubility.

Separation Scheme

The gradient elution scheme is a scaled-up procedure originally described by Middleton (10) that has been extended to handle highly refractive materials such as coal liquids. This separation technique uses Alcoa F-20 alumina activated to a 5.5 wt % moisture level as the stationary phase. Details of this separation procedure are given elsewhere (2). This method separates SRC into 13 fractions and these fractions are listed in Table II along with some key chemical and physical descriptions of the cuts. The structural types indicated in Table II for Fractions 1–6 have been assigned based upon model compound studies and low resolution mass spectrometry (MS) (2), whereas the chemical types indicated for Fractions 7–13 are based upon IR observations and additional model compound studies. Recoveries in these separations are normally greater than 90%.

Table II.   Chemical and Physical Nature

GEC Fraction No.	Fraction Name
1	saturates
2	MNA–DNA oils
3	PNA oils
4	PNA soft resin
5	hard resin
6	polar resin
7	asphaltenes
8–12	polar asphaltenes
13	non-eluted

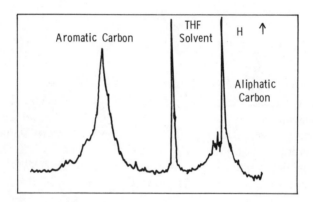

Figure 1a.   C-13 NMR spectrum of Burning Star SRC

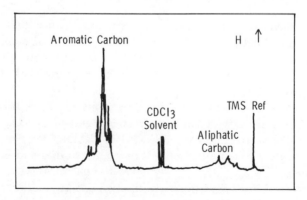

Figure 1b.   C-B NMR spectrum of the PNA soft resin fraction from SRC

of Gradient Elution Fractions

Chemical Type	Physical Description
paraffins	colorless oil or white wax
cycloparaffins	
mono + dinuclear	very light yellow oil
aromatic hydrocarbons	
polynuclear	bright yellow/orange oil
aromatic hydrocarbons	
O, S heterocyclics	tacky orange–red semi-solid,
polynuclear aromatics	fluid at 200°F
O, S, N heterocyclics	red–brown powder
O, S, N heterocyclics	red–brown powder
phenolic functional groups	dark brown powder
multifunctional compounds	black powder
unreacted coal, ash,	
strong acids	

NMR Studies

Both C-13 NMR and proton-NMR were obtained by the Fourier transform technique using a Varian Model CFT-20 spectrometer. For C-13 work a quantitative representation of the different types of carbon atoms was obtained by operating in the gated decoupling mode to minimize the bias caused by nuclear Overhauser enhancement (NOE). A pulse delay of 25 sec with a flip angle of 90° was sufficient to assure reasonably complete relaxation of most carbon atoms. Delays greater than 25 sec did not yield any measurable differences in the area integration for aliphatic and aromatic carbon atoms. In most cases samples were run in solutions of deutero-chloroform (used for internal signal lock). However, the poor solubility in this solvent of whole SRC and of the latter eluting GEC fractions required the use of THF as a solvent. The latter necessitated the use of D_2O placed in an axially centered capillary within the sample tube to provide a sample lock signal. Interfering solvent peaks caused by THF were integrated and subtracted from the spectrum for the calculation of aromaticities. Figure 1 shows two typical examples of C-13 NMR spectra obtained in this work. The high signal/noise ratios inherent in these spectra enable aromaticities to be measured with a repeatability of $\pm 2\%$. Aromaticity values are expressed as the fraction, f_a, of aromatic carbon atoms relative to the total of both aliphatic and aromatic carbon atoms. Olefinic carbon atoms are indistinguishable from aromatic carbons and, if present, would be included with aromatic carbon atoms.

Table III. Elemental Analyses, Molecular Weights, and Aromaticities of SRC's and Vacuum Resid

	Illinois No. 6 Burning Star SRC	Illinois No. 6 Monterey SRC	Wyodak Amax SRC	Arabian Light Vacuum Resid
Carbon	87.9	85.4	87.7	84.6
Hydrogen	5.7	5.8	5.6	10.5
Oxygen	3.5	4.3	3.8	0.7
Nitrogen	1.7	2.1	1.7	0.3
Sulfur	0.6	1.0	0.1	3.8
Molecular weight	519	616	632	1020
Aromaticity, f_a	0.77	0.73	0.79	0.30

Results and Discussion

Elemental analyses of the three SRC's and the petroleum resid along with number average molecular weights and aromaticities as determined by C-13 NMR are presented in Table III. In general, all three SRC's are higher in both oxygen and nitrogen content than the vacuum resid and lower in sulfur and hydrogen. The higher carbon/hydrogen ratios of the SRC's, as compared with the vacuum resid, is reflected in the higher aromaticities observed for these coal liquids. Indeed, the C-13 NMR data demonstrate that SRC's from these coal sources are highly aromatic materials.

Table IV.   GEC Analyses of SRC Products + Vacuum Resid

Cut	Fraction	Illinois No. 6 Burning Star (Wt %)	Illinois No. 6 Monterey (Wt %)	Wyodak Amax (Wt %)	Arabian Light Vacuum Resid (Wt %)
1	Saturates	0.04	0.02	0.03	11.0
2	MNA–DNA oil	0.2	1.1	0.2	18.6
3	PNA oil	0.9	1.8	2.6	19.7
4	PNA soft resin	10.8	4.8	6.3	28.0
5	Hard resin	2.4	2.4	2.4	4.5
6	Polar resin	4.5	4.9	4.0	6.1
7	Asphaltenes	23.9	25.7	22.4	8.9
8	"Polar" asphaltenes	14.6	15.9	15.6	2.3
9	"Polar" asphaltenes	6.1	5.5	6.8	0.2
10	"Polar" asphaltenes	5.9	11.0	7.1	0.04
11	"Polar" asphaltenes	13.9	17.7	17.0	0.06
12	"Polar" asphaltenes	6.7	4.0	4.5	0.02
13	Non-eluted + loss	10.06	5.18	11.07	0.58

Comparison of the elemental composition of the SRC's shows that Illinois No. 6 Monterey SRC is quite different from the SRC derived from Illinois No. 6 Burning Star Coal. The Monterey sample is higher in oxygen, nitrogen, and sulfur than the Burning Star SRC. By contrast the Amax SRC derived from the Wyodak seam coal contains about the same amount of nitrogen and oxygen as the Burning Star sample, but significantly less sulfur than either of the Illinois SRC's.

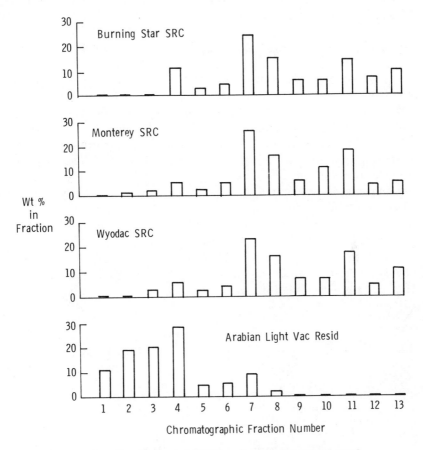

Figure 2. Comparison of SRC products and vacuum resid

The GEC analyses of the SRC's are presented in Table IV and compared with a comparable analysis for an Arabian Light vacuum resid. This comparison is illustrated graphically in Figure 2. The most striking feature of this data lies in the observed difference in composition between the SRC's and the resid. Whereas the Arabian Light vacuum resid contains ~ 49 wt % material in the first three fractions, the SRC's contain less than 3 wt % in these categories. On the other hand, more than 80 wt % of the SRC's elutes in Fractions 7–13 compared with only 12 wt % for the resid. Subtle composition differences also were observed between the various SRC materials. Thus, compared with Burning Star SRC, Monterey SRC contains more aromatic oils (Fractions 2 and 3) but less of the soft resin fraction (Fraction 4); Monterey SRC also contains significantly more of Polar Fractions 10 and 11, but a great deal less of Cuts 12 and 13. Wyodak Amax SRC appears to be intermediate between the two SRC's.

Characterization of GEC Fractions

Considerable attention has been devoted to characterizing the gradient elution fractions in structural terms. Carbon and hydrogen analyses on the GEC fractions from the three SRC's are presented in Table 5. As one progresses from the earlier fractions to the more polar fractions there is a gradual increase in the carbon/hydrogen atom ratio suggesting that the latter fractions are more aromatic. The heteroatom concentrations of the GEC fractions are shown in Figures 3, 4, and 5 for the Burning Star, Monterey, and Wyodak SRC, respectively. All three SRC liquids show similar trends in their heteroatom concentration profiles. The nitrogen concentration rises sharply at Fraction 5, reaches a maximum in Fraction 7 and then drops slightly to a level that is fairly constant throughout Fractions 8–12. The oxygen level increases dramatically in Fraction 7 and then rises gradually through Fractions 8 to 12. IR spectra of Fractions 7–12 show the presence of a broad band centered

Table V.   Carbon and Hydrogen Analyses

		Burning Star SRC	
Cut No.	GEC Fraction	C (Wt %)	H (Wt %)
1	Saturates	[a]	[a]
2	MNA + DNA oil	[a]	[a]
3	PNA oil	88.4	7.9
4	PNA soft resin	91.8	6.0
5	Hard resin	90.7	6.1
6	Polar resin	89.0	6.1
7	Eluted asphaltenes	87.2	6.2
8	Polar asphaltenes	86.0	7.0
9	Polar asphaltenes	87.1	5.5
10	Polar asphaltenes	87.1	5.4
11	Polar asphaltenes	85.0	5.6
12	Polar asphaltenes	83.3	5.4

[a] Analysis not determined.

at 3300 cm⁻¹ and a second band at 3530 cm⁻¹. The latter band is attributed to the presence of free OH and N–H functional groups. Although the presence of phenolic groups in these fractions is a certainty, the ratio of phenolic oxygen to total oxygen has not been determined. However, retention volume considerations based on measurements with model compounds indicate that oxygen present in Fractions 2, 3, and 4 can only be cyclic ethers present in either aromatic ring systems

or as ether bridges.

of GEC Fractions from Wilsonville SRC's

Monterey SRC		Wyodak SRC	
C (Wt %)	H (Wt %)	C (Wt %)	H (Wt %)
a	a	a	a
89.4	8.8	88.9	11.2
90.7	6.7	91.8	6.9
90.6	6.5	91.8	6.3
89.2	6.1	91.3	5.9
89.2	6.2	91.2	5.9
84.1	6.7	85.6	6.6
83.8	6.4	85.5	6.4
84.1	5.7	86.7	5.7
83.8	5.8	86.5	5.3
80.7	6.1	84.3	5.9
83.6	6.0	83.7	5.3

Sulfur is fairly evenly distributed throughout the GEC fractions for all three SRC's. Low-resolution MS of Fraction 4 from Burning Star SRC indicated significant concentrations of the aromatic sulfur types such as benzothiophenes, dibenzothiophenes, and naphthobenzothiophenes. The occurrence of these sulfur compounds in coal liquids has been observed before (11) and their presence in SRC is not unexpected. Sulfur in Fractions 7–12 must occur in molecules that also contain nitrogen and/or oxygen functional groups.

The heteroatom profiles shown for these SRC's in Figures 3, 4, and 5 provide a useful macroscopic characterization of the coal liquid components within a given chromatographic fraction. This information when coupled with comparable data on hydroprocessed coal liquids can yield valuable insights into the nature of the heteroatom removal reactions (3).

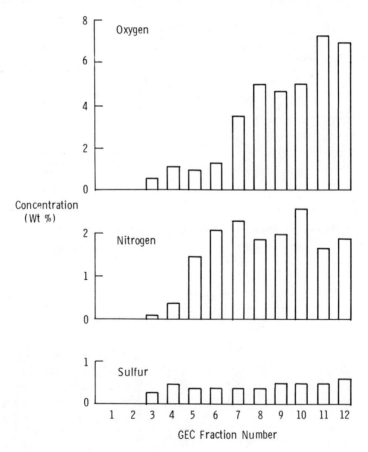

Figure 3.   Heteroatom concentrations in Burning Star SRC GEC fractions

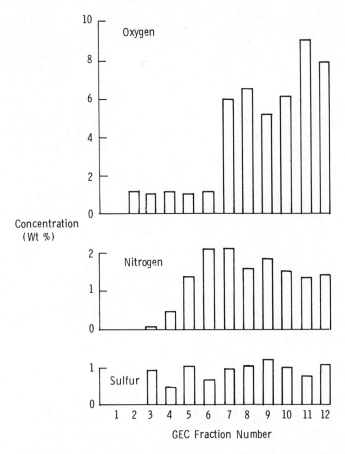

*Figure 4.   Heteroatom concentrations in Monterey SRC
GEC fractions*

Table VI.   Molecular Weights of Chromatographic Fractions

Cut No.	GEC Fraction	Burning Star SRC	Monterey SRC	Wyodak SRC	Arabian Light Vacuum Resid
4	PNA soft resin	301	*a*	352	940
7	asphaltenes	478	580	590	1571
8	polar asphaltenes	587	736	835	5900
11	polar asphaltenes	645	905	939	*a*

a Analysis not determined.

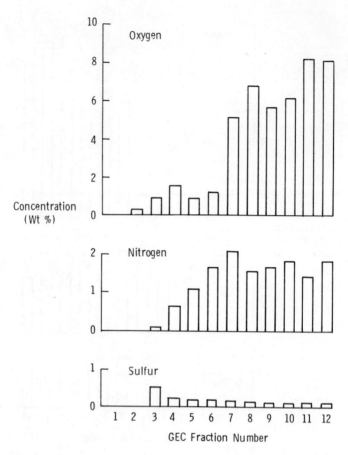

*Figure 5. Heteroatom concentrations in Amax SRC GEC
fractions*

Another interesting contrast between SRC's and petroleum residua
is seen by comparing the molecular weights of the SRC GEC fractions
with the measured molecular weights on Arabian Light vacuum resid
fractions. The data in Table VI show that the SRC fractions are consistently lower in molecular weight than the resid fractions. The molecular
weight measurements were all performed at equivalent concentrations
(ca. 1%) of solute in THF and consequently molecular association is not
believed to be a significant contributing factor to these observed
differences. These data thus indicate that the molecular sizes of components present in coal liquids, at the coal conversion levels presented
in Table I, are significantly smaller than components in petroleum
residua.

NMR Studies

Nuclear magnetic resonance (NMR) spectroscopy is a powerful tool for obtaining structural parameters of interest for processing both coal liquids and petroleum residua. Aromaticity, f_a, has long been recognized as a key parameter for characterizing petroleum fractions (12, 13) and several methods have been proposed for determining aromaticity from elemental analyses and H-NMR data (14, 15, 16). On the other hand C-13 NMR allows a direct measurement of aromaticity and recent instrumentation advances have enabled reliable C-13 NMR measurements to be made at natural abundance levels in reasonable amounts of time. We have used C-13 NMR to measure aromaticities of selected GEC fractions from the SRC's examined in this study and Table VII contains our findings. The data in Table VII show that all SRC fractions are highly aromatic with f_a ranging from 0.75 to 0.91. Our data also show that the more polar fractions from SRC are slightly more aromatic than the earlier eluting fractions. However, the most striking feature about the data in Table VII is the contrast between the aromaticity of the SRC fractions and the aromaticity of the fractions obtained from Arabian Light vacuum resid. Thus, Fraction 4 from the petroleum resid has an f_a of 0.42 compared with 0.83 for this same fraction isolated from Burning Star SRC, and 0.79 for the comparable fractions obtained from both Monterey and Amax SRC's. In similar fashion, Fraction 8 from the petroleum resid has an f_a of 0.52 in contrast to corresponding values of 0.80, 0.84, and 0.82 for Fraction 8 obtained from Burning Star, Monterey, and Amax SRC. One structural feature that showed a substantial difference between the petroleum resid and SRC's is the average substituent length as calculated from C-13 and H-NMR measurements. This difference is illustrated by data in Table VIII, wherein the average chain length for Fractions 4 and 8 obtained from Monterey SRC and Arabian Light vacuum resid are compared. For the SRC fractions, the average chain length is less than 2 indicating a predominance of methyl, ethyl, and hydroaromatic

Table VII.   Aromaticities, f_a, of Selected GEC Fractions

Fraction No.	GEC Fraction	Burning Star SRC	Monterey SRC	Wyodak Amax SRC	Arabian Light Vacuum Resid
4	PNA soft resin	0.83	0.79	0.79	0.42
7	asphaltenes	0.79	0.75	0.76	[a]
8	polar asphaltenes	0.80	0.84	0.82	0.52
11	polar asphaltenes	0.91	0.82	0.75	[a]

[a] Analysis not determined.

**Table VIII.   Structural Properties of Selected GEC Fractions
from SRC vs. Vacuum Resid**

	Sample			
	Monterey SRC	*Arab Light Vacuum Resid*	*Monterey SRC*	*Arab Light Vacuum Resid*
Fraction No.	4	4	8	8
Aromaticity, f_a	0.79	0.42	0.84	0.52
H_{ar}	43	8	44	16
H_α	41	24	36	22
H_o	16	68	20	62
Average chain length	1.4	3.8	1.6	5.4

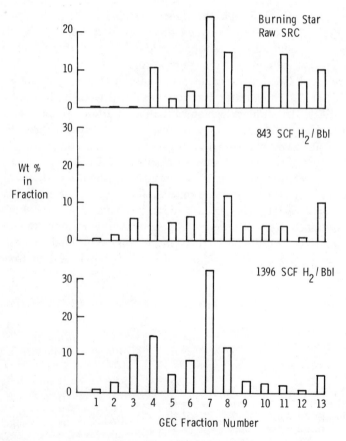

Figure 6.   GEC comparison of raw and hydroprocessed SRC

substituents. On the other hand, the average chain length for the resid fractions is significantly higher indicating both longer chains and poly-naphthenic substituents.

Conclusion

This research has demonstrated that GEC is a valuable separation tool for characterizing SRC and comparing the composition of these coal liquids with petroleum-derived material with comparable boiling range distributions. The SRC's are richer in nitrogen and oxygen functionalities than the residua, but the molecular species in the coal liquids are significantly lower in molecular weight. The NMR results on these SRC's indicate that these materials are much more highly aromatic than residua and that there is a substantial difference in the nature of the alkyl substituents on the aromatic moieties. Finally, as these coal liquids are upgraded by hydroprocessing, aromaticity and heteroatom content both are reduced substantially and in general the upgraded coal liquids become more petroleum-like. The composition changes that occur through hydro-processing also can be followed conveniently by the gradient elution technique used here. Figure 6 illustrates this feature wherein Burning Star SRC is compared with two hydroprocessed SRC products. As hydro-processing progresses, there is an overall conversion of Fractions 8–13 resulting in a net increase in the saturate, aromatic oil, and resin fractions.

Acknowledgment

The work on SRC is conducted under Electric Power Research Institute (EPRI) Contract No. RP 361, which is jointly funded by EPRI and Mobil Research and Development Corporation. W. C. Rovesti is EPRI project manager.

Literature Cited

1. Bodle, W. W., Vyas, K. C., *Oil Gas J.* (1974) **August 26, 73.**
2. Callen, R. B., Bendoraitis, J. G., Simpson, C. A., Voltz, S. E., *I & EC Prod. Res. Dev.* (1976) **15,** 222.
3. Stein, T. R., Voltz, S. E., Callen, R. B., *I & EC Prod. Res. Dev.* (1977) **16,** 61.
4. Sternberg, H. W., Raymond, R., Schweighardt, F. K., *Prepr., Div. Pet. Chem., Am. Chem. Soc.* (1976) **21**(1), 198.
5. Schwager, I., Yen, T. F., *Am. Chem. Soc., Div. Fuel Chem., Prepr.* (1976) **21**(5), 199.
6. Ruberto, R. G., Jewell, D. M., Jensen, R. K., Cronauer, D. C., Am. Chem. Soc., *Div. Fuel Chem., Prepr.* (1974) **19**(2), 258.
7. Farcasiu, M., *Fuel* (1977) **56,** 9.
8. Farcasiu, M., Mitchell, T. O., Whitehurst, D. D., 1st Yearly Report **RP-410-1** to EPRI, February (1976).

9. Sternberg, H. W., Raymond, R., Schweighardt, F. K., *Science* (1975) **188**, 49.
10. Middleton, W. R., *Anal. Chem.* (1967) **39**, 1839.
11. Akhtar, S., Sharkey, A. G., Schultz, J. L., Yavorsky, P. M., *Am. Chem. Soc., Div. of Fuel Chem., Prepr.* (1974) **19**(1), 207.
12. Knight, S. A., Jenkins, G. I., *Chem. Ind. London* (1972) **1972**, 614.
13. Williams, R. B., Chamberlain, N. F., *World Pet.* (World Pet. Congr., Proc., 6th) (1963) **34**(6), 217.
14. Brown, J. K., Ladner, W. R., *Fuel* (1967) **39**, 87.
15. Speight, J. G., *Fuel* (1970) **49**, 76.
16. Clutter, D. R., Petrakis, L., Stenger, R. L., Jensen, R. K., *Anal. Chem.* (1972) **44**, 1395.

RECEIVED October 17, 1977.

INDEX

INDEX

The text of this book is set in 10 point Caledonia with two points of leading. The chapter numerals are set in 30 point Garamond; the chapter titles are set in 18 point Garamond Bold.

The book is printed offset on Text White Opaque 50-pound
The cover is Joanna Book Binding blue linen.

Jacket design by Alan Kahan.
Editing and production by Candace A. Deren.

The book was composed by Service Composition Co., Baltimore, Md., printed and bound by The Maple Press Co., York, PA.